Probability and Statistics
For Engineers and Scientists

FOURTH EDITION

Anthony J. Hayter
Denver University

BROOKS/COLE
CENGAGE Learning·

Australia • Brazil • Japan • Korea • Mexico • Singapore • Spain • United Kingdom • United States

ISBN-13: 978-1-133-11131-3
ISBN-10: 1-133-11131-9

Brooks/Cole
20 Channel Center Street
Boston, MA 02210
USA

Cengage Learning is a leading provider of customized learning solutions with office locations around the globe, including Singapore, the United Kingdom, Australia, Mexico, Brazil, and Japan. Locate your local office at: **www.cengage.com/global**

Cengage Learning products are represented in Canada by Nelson Education, Ltd.

To learn more about Brooks/Cole, visit **www.cengage.com/brookscole**

Purchase any of our products at your local college store or at our preferred online store **www.cengagebrain.com**

Printed in the United States of America
1 2 3 4 5 6 7 15 14 13 12 11

Student Solution Manual

This student solution manual to accompany the fourth edition of

"Probability and Statistics for Engineers and Scientists" by Anthony Hayter

provides worked solutions and answers to the odd-numbered problems given at the end of the chapter sections.

Contents

Chapter 1

Probability Theory

1.1 Probabilities

1.1.1 $\mathcal{S} = \{$(head, head, head), (head, head, tail), (head, tail, head),

        (head, tail, tail), (tail, head, head), (tail, head, tail),

        (tail, tail, head), (tail, tail, tail)$\}$

1.1.3 $\mathcal{S} = \{0,1,2,3,4\}$

1.1.5 $\mathcal{S} = \{$(on time, satisfactory), (on time, unsatisfactory),

        (late, satisfactory), (late, unsatisfactory)$\}$

1.1.7   (a) $\frac{p}{1-p} = 1 \;\Rightarrow\; p = 0.5$

       (b) $\frac{p}{1-p} = 2 \;\Rightarrow\; p = \frac{2}{3}$

       (c) $p = 0.25 \;\Rightarrow\; \frac{p}{1-p} = \frac{1}{3}$

1.1.9 $0.08 + 0.20 + 0.33 + P(IV) + P(V) = 1 \;\Rightarrow\; P(IV) + P(V) = 1 - 0.61 = 0.39$

       Therefore, $0 \le P(V) \le 0.39$.

       If $P(IV) = P(V)$ then $P(V) = 0.195$.

1.1.11 $p = 1 - 0.28 - 0.55 = 0.17$.

1.2   Events

1.2.1   (a) $0.13 + P(b) + 0.48 + 0.02 + 0.22 = 1 \Rightarrow P(b) = 0.15$

     (b) $A = \{c, d\}$ so that $P(A) = P(c) + P(d) = 0.48 + 0.02 = 0.50$

     (c) $P(A') = 1 - P(A) = 1 - 0.5 = 0.50$

1.2.3   Over a four year period including one leap year, the number of days is

$(3 \times 365) + 366 = 1461$.

The number of January days is $4 \times 31 = 124$

and the number of February days is $(3 \times 28) + 29 = 113$.

The answers are therefore $\frac{124}{1461}$ and $\frac{113}{1461}$.

1.2.5 $1 - 0.38 - 0.11 - 0.16 = 0.35$

$0.38 + 0.16 + 0.35 = 0.89$

1.2.7 $P(\spadesuit \text{ or } \clubsuit) = P(A\spadesuit) + P(K\spadesuit) + \ldots + P(2\spadesuit) + P(A\clubsuit) + P(K\clubsuit) + \ldots + P(2\clubsuit) = \frac{1}{52} + \ldots + \frac{1}{52} = \frac{26}{52} = \frac{1}{2}$

1.2.9   (a) Let the four players be named A, B, C, and T for Terica, and let the notation (X, Y) indicate that player X is the winner and player Y is the runner up.

The sample space consists of the 12 outcomes:
$\mathcal{S} = \{$(A,B), (A,C), (A,T), (B,A), (B,C), (B,T), (C,A), (C,B), (C,T), (T,A), (T,B), (T,C)$\}$

The event *'Terica is winner'* consists of the 3 outcomes $\{$(T,A), (T,B), (T,C)$\}$.

Since each outcome in $\mathcal{S}$ is equally likely to occur with a probability of $\frac{1}{12}$ it follows that
$P(\text{Terica is winner}) = \frac{3}{12} = \frac{1}{4}$.

(b) The event *'Terica is winner or runner up'* consists of 6 out of the 12 outcomes so that

P(Terica is winner or runner up) $= \frac{6}{12} = \frac{1}{2}$.

1.2.11 (a) See Figure 1.25.

The event *'both assembly lines are shut down'* consists of the single outcome $\{(S,S)\}$.

Therefore,

P(both assembly lines are shut down) $= 0.02$.

(b) The event *'neither assembly line is shut down'* consists of the outcomes

$\{(P,P), (P,F), (F,P), (F,F)\}$.

Therefore,

P(neither assembly line is shut down)
$= P((P,P)) + P((P,F)) + P((F,P)) + P((F,F))$
$= 0.14 + 0.2 + 0.21 + 0.19 = 0.74$.

(c) The event *'at least one assembly line is at full capacity'* consists of the outcomes

$\{(S,F), (P,F), (F,F), (F,S), (F,P)\}$.

Therefore,

P(at least one assembly line is at full capacity)
$= P((S,F)) + P((P,F)) + P((F,F)) + P((F,S)) + P((F,P))$
$= 0.05 + 0.2 + 0.19 + 0.06 + 0.21 = 0.71$.

(d) The event *'exactly one assembly line at full capacity'* consists of the outcomes

$\{(S,F), (P,F), (F,S), (F,P)\}$.

Therefore,

P(exactly one assembly line at full capacity)
$= P((S,F)) + P((P,F)) + P((F,S)) + P((F,P))$
$= 0.05 + 0.20 + 0.06 + 0.21 = 0.52$.

The complement of *'neither assembly line is shut down'* is the event *'at least one assembly line is shut down'* which consists of the outcomes

{(S,S), (S,P), (S,F), (P,S), (F,S)}.

The complement of *'at least one assembly line is at full capacity'* is the event *'neither assembly line is at full capacity'* which consists of the outcomes

{(S,S), (S,P), (P,S), (P,P)}.

1.2.13 $0.26 + 0.36 + 0.11 = 0.73$

1.3 Combinations of Events

1.3.1   The event A contains the outcome 0 while the empty set does not contain any outcomes.

1.3.5   Yes, because a card must be drawn from either a red suit or a black suit but it cannot be from both at the same time.

No, because the ace of hearts could be drawn.

1.3.7   Since $P(A \cup B) = P(A) + P(B) - P(A \cap B)$

it follows that

$P(B) = P(A \cup B) - P(A) + P(A \cap B)$

$= 0.8 - 0.5 + 0.1 = 0.4.$

1.3.9   Yes, the three events are mutually exclusive because the selected card can only be from one suit.

Therefore,

$P(A \cup B \cup C) = P(A) + P(B) + P(C) = \frac{1}{4} + \frac{1}{4} + \frac{1}{4} = \frac{3}{4}.$

A' is the event *'a heart is not obtained'* (or similarly the event *'a club, spade, or diamond is obtained'*) so that B is a subset of A'.

1.3.11   Let the event O be an on time repair and let the event S be a satisfactory repair.

It is known that $P(O \cap S) = 0.26$, $P(O) = 0.74$ and $P(S) = 0.41$.

We want to find $P(O' \cap S')$.

Since the event $O' \cap S'$ can be written $(O \cup S)'$ it follows that

$P(O' \cap S') = 1 - P(O \cup S)$

$= 1 - (P(O) + P(S) - P(O \cap S))$

$= 1 - (0.74 + 0.41 - 0.26) = 0.11.$

1.3.13   Let A be the event that the patient is male, let B be the event that the patient is younger than thirty years of age, and let C be the event that the patient is admitted to the hospital.

It is given that $P(A) = 0.45$, $P(B) = 0.30$, $P(A' \cap B' \cap C) = 0.15$, and $P(A' \cap B) = 0.21$.

The question asks for $P(A' \cap B' \cap C')$.

Notice that

$P(A' \cap B') = P(A') - P(A' \cap B) = (1 - 0.45) - 0.21 = 0.34$

so that

$P(A' \cap B' \cap C') = P(A' \cap B') - P(A' \cap B' \cap C) = 0.34 - 0.15 = 0.19.$

1.3.15 $P(A \cap B) = P(B) = 0.43 + 0.29 = 0.72$

$P(A \cup B) = P(A) = 0.18 + 0.43 + 0.29 = 0.90$

1.4   Conditional Probability

1.4.1   See Figure 1.55.

(a) $P(A \mid B) = \frac{P(A \cap B)}{P(B)} = \frac{0.02+0.05+0.01}{0.02+0.05+0.01+0.11+0.08+0.06+0.13} = 0.1739$

(b) $P(C \mid A) = \frac{P(A \cap C)}{P(A)} = \frac{0.02+0.05+0.08+0.04}{0.02+0.05+0.08+0.04+0.018+0.07+0.05} = 0.59375$

(c) $P(B \mid A \cap B) = \frac{P(B \cap (A \cap B))}{P(A \cap B)} = \frac{P(A \cap B)}{P(A \cap B)} = 1$

(d) $P(B \mid A \cup B) = \frac{P(B \cap (A \cup B))}{P(A \cup B)} = \frac{P(B)}{P(A \cup B)} = \frac{0.46}{0.46+0.32-0.08} = 0.657$

(e) $P(A \mid A \cup B \cup C) = \frac{P(A \cap (A \cup B \cup C))}{P(A \cup B \cup C)} = \frac{P(A)}{P(A \cup B \cup C)} = \frac{0.32}{1-0.04-0.05-0.03}$
$= 0.3636$

(f) $P(A \cap B \mid A \cup B) = \frac{P((A \cap B) \cap (A \cup B))}{P(A \cup B)} = \frac{P(A \cap B)}{P(A \cup B)} = \frac{0.08}{0.7} = 0.1143$

1.4.3   (a) $P(A\heartsuit \mid \text{red suit}) = \frac{P(A\heartsuit \cap \text{red suit})}{P(\text{red suit})} = \frac{P(A\heartsuit)}{P(\text{red suit})} = \frac{\left(\frac{1}{52}\right)}{\left(\frac{26}{52}\right)} = \frac{1}{26}$

(b) $P(\text{heart} \mid \text{red suit}) = \frac{P(\text{heart} \cap \text{red suit})}{P(\text{red suit})} = \frac{P(\text{heart})}{P(\text{red suit})} = \frac{\left(\frac{13}{52}\right)}{\left(\frac{26}{52}\right)}$
$= \frac{13}{26} = \frac{1}{2}$

(c) $P(\text{red suit} \mid \text{heart}) = \frac{P(\text{red suit} \cap \text{heart})}{P(\text{heart})} = \frac{P(\text{heart})}{P(\text{heart})} = 1$

(d) $P(\text{heart} \mid \text{black suit}) = \frac{P(\text{heart} \cap \text{black suit})}{P(\text{black suit})} = \frac{P(\emptyset)}{P(\text{black suit})} = 0$

(e) $P(\text{King} \mid \text{red suit}) = \frac{P(\text{King} \cap \text{red suit})}{P(\text{red suit})} = \frac{P(K\heartsuit, K\diamondsuit)}{P(\text{red suit})} = \frac{\left(\frac{2}{52}\right)}{\left(\frac{26}{52}\right)} = \frac{2}{26}$
$= \frac{1}{13}$

(f) $P(\text{King} \mid \text{red picture card}) = \frac{P(\text{King} \cap \text{red picture card})}{P(\text{red picture card})}$

$$= \frac{P(K\heartsuit,\, K\diamondsuit)}{P(\text{red picture card})} = \frac{\left(\frac{2}{52}\right)}{\left(\frac{6}{52}\right)} = \frac{2}{6} = \frac{1}{3}$$

1.4.5   There are 54 blue balls and so there are $150 - 54 = 96$ red balls.

Also, there are 36 shiny, red balls and so there are $96 - 36 = 60$ dull, red balls.

$$P(\text{shiny} \mid \text{red}) = \frac{P(\text{shiny} \cap \text{red})}{P(\text{red})} = \frac{\left(\frac{36}{150}\right)}{\left(\frac{96}{150}\right)} = \frac{36}{96} = \frac{3}{8}$$

$$P(\text{dull} \mid \text{red}) = \frac{P(\text{dull} \cap \text{red})}{P(\text{red})} = \frac{\left(\frac{60}{150}\right)}{\left(\frac{96}{150}\right)} = \frac{60}{96} = \frac{5}{8}$$

1.4.7   (a) It depends on the weather patterns in the particular location that is being considered.

    (b) It increases since there are proportionally more black haired people among brown eyed people than there are in the general population.

    (c) It remains unchanged.

    (d) It increases.

1.4.9   (a) Let A be the event that *'Type I battery lasts longest'* consisting of the outcomes $\{(\text{III, II, I}), (\text{II, III, I})\}$.

Let B be the event that *'Type I battery does not fail first'* consisting of the outcomes $\{(\text{III,II,I}), (\text{II,III,I}), (\text{II,I,III}), (\text{III,I,II})\}$.

The event $A \cap B = \{(\text{III,II,I}), (\text{II,III,I})\}$ is the same as event A.

Therefore,

$$P(A \mid B) = \frac{P(A \cap B)}{P(B)} = \frac{0.39 + 0.03}{0.39 + 0.03 + 0.24 + 0.16} = 0.512.$$

    (b) Let C be the event that *'Type II battery fails first'* consisting of the outcomes $\{(\text{II,I,III}), (\text{II,III,I})\}$.

Thus, $A \cap C = \{(II, III, I)\}$ and therefore

$$P(A \mid C) = \frac{P(A \cap C)}{P(C)} = \frac{0.39}{0.39 + 0.24} = 0.619.$$

(c) Let D be the event that *'Type II battery lasts longest'* consisting of the outcomes $\{(I,III,II), (III,I,II)\}$.

Thus, $A \cap D = \emptyset$ and therefore

$P(A \mid D) = \frac{P(A \cap D)}{P(D)} = 0.$

(d) Let E be the event that *'Type II battery does not fail first'* consisting of the outcomes $\{(I,III,II), (I,II,III), (III,II,I), (III,I,II)\}$.

Thus, $A \cap E = \{(III, II, I)\}$ and therefore

$P(A \mid E) = \frac{P(A \cap E)}{P(E)} = \frac{0.03}{0.07+0.11+0.03+0.16} = 0.081.$

1.4.11 Let L, W and H be the events that the length, width and height respectively are within the specified tolerance limits.

It is given that $P(W) = 0.86$, $P(L \cap W \cap H) = 0.80$, $P(L \cap W \cap H') = 0.02$, $P(L' \cap W \cap H) = 0.03$ and $P(W \cup H) = 0.92$.

(a) $P(W \cap H) = P(L \cap W \cap H) + P(L' \cap W \cap H) = 0.80 + 0.03 = 0.83$

$P(H) = P(W \cup H) - P(W) + P(W \cap H) = 0.92 - 0.86 + 0.83 = 0.89$

$P(W \cap H \mid H) = \frac{P(W \cap H)}{P(H)} = \frac{0.83}{0.89} = 0.9326$

(b) $P(L \cap W) = P(L \cap W \cap H) + P(L \cap W \cap H') = 0.80 + 0.02 = 0.82$

$P(L \cap W \cap H \mid L \cap W) = \frac{P(L \cap W \cap H)}{P(L \cap W)} = \frac{0.80}{0.82} = 0.9756$

1.4.13 (a) Let E be the event that the *'component passes on performance'*, let A be the event that the *'component passes on appearance'*, and let C be the event that the *'component passes on cost'*.

$P(A \cap C) = 0.40$

$P(E \cap A \cap C) = 0.31$

$P(E) = 0.64$

$P(E' \cap A' \cap C') = 0.19$

$P(E' \cap A \cap C') = 0.06$

Therefore,

$P(E' \cap A' \cap C) = P(E' \cap A') - P(E' \cap A' \cap C')$

$= P(E') - P(E' \cap A) - 0.19$

$= 1 - P(E) - P(E' \cap A \cap C) - P(E' \cap A \cap C') - 0.19$

$$= 1 - 0.64 - P(A \cap C) + P(E \cap A \cap C) - 0.06 - 0.19$$
$$= 1 - 0.64 - 0.40 + 0.31 - 0.06 - 0.19 = 0.02$$

(b) $P(E \cap A \cap C \mid A \cap C) = \frac{P(E \cap A \cap C)}{P(A \cap C)}$

$= \frac{0.31}{0.40} = 0.775$

1.4.15 $P(\text{delay}) = (P(\text{delay} \mid \text{technical problems}) \times P(\text{technical problems}))$

$+ (P(\text{delay} \mid \text{no technical problems}) \times P(\text{no technical problems}))$

$= (1 \times 0.04) + (0.33 \times 0.96) = 0.3568$

1.4.17 $\frac{0.26}{0.26 + 0.36 + 0.11} = 0.3562$

1.5 Probabilities of Event Intersections

1.5.1   (a) $P(\text{both cards are picture cards}) = \frac{12}{52} \times \frac{11}{51} = \frac{132}{2652}$

(b) $P(\text{both cards are from red suits}) = \frac{26}{52} \times \frac{25}{51} = \frac{650}{2652}$

(c) $P(\text{one card is from a red suit and one is from black suit})$
$= (P(\text{first card is red}) \times P(\text{2nd card is black} \mid \text{1st card is red}))$
$+ (P(\text{first card is black}) \times P(\text{2nd card is red} \mid \text{1st card is black}))$
$= \left(\frac{26}{52} \times \frac{26}{51}\right) + \left(\frac{26}{52} \times \frac{26}{51}\right) = \frac{676}{2652} \times 2 = \frac{26}{51}$

1.5.3   (a) No, they are not independent.
Notice that
$P((ii)) = \frac{3}{13} \neq P((ii) \mid (i)) = \frac{11}{51}$.

(b) Yes, they are independent.
Notice that
$P((i) \cap (ii)) = P((i)) \times P((ii))$
since
$P((i)) = \frac{1}{4}$
$P((ii)) = \frac{3}{13}$
and
$P((i) \cap (ii)) = P(\text{first card a heart picture} \cap (ii))$
$+ P(\text{first card a heart but not a picture} \cap (ii))$
$= \left(\frac{3}{52} \times \frac{11}{51}\right) + \left(\frac{10}{52} \times \frac{12}{51}\right) = \frac{153}{2652} = \frac{3}{52}$.

(c) No, they are not independent.
Notice that
$P((ii)) = \frac{1}{2} \neq P((ii) \mid (i)) = \frac{25}{51}$.

(d) Yes, they are independent.
Similar to part (b).

(e) No, they are not independent.

1.5.5 $P(\text{all 4 cards are hearts}) = (\frac{13}{52})^4 = \frac{1}{256}$

The probability increases with replacement.

$P(\text{all 4 cards are from red suits}) = (\frac{26}{52})^4 = \frac{1}{16}$

The probability increases with replacement.

$P(\text{all 4 cards from different suits}) = 1 \times \frac{39}{52} \times \frac{26}{52} \times \frac{13}{52} = \frac{3}{32}$

The probability decreases with replacement.

1.5.7   The only way that a message will not get through the network is if both
branches are closed at the same time. The branches are independent since
the switches operate independently of each other.

Therefore,

$P(\text{message gets through the network})$

$= 1 - P(\text{message cannot get through the top branch or the bottom branch})$

$= 1 - (P(\text{message cannot get through the top branch})$

$\times P(\text{message cannot get through the bottom branch}))$.

Also,

$P(\text{message gets through the top branch}) = P(\text{switch 1 is open} \cap \text{switch 2 is open})$

$= P(\text{switch 1 is open}) \times P(\text{switch 2 is open})$

$= 0.88 \times 0.92 = 0.8096$

since the switches operate independently of each other.

Therefore,

$P(\text{message cannot get through the top branch})$

$= 1 - P(\text{message gets through the top branch})$

$= 1 - 0.8096 = 0.1904$.

Furthermore,

$P(\text{message cannot get through the bottom branch})$

$= P(\text{switch 3 is closed}) = 1 - 0.9 = 0.1$.

Therefore,

$P(\text{message gets through the network}) = 1 - (0.1 \times 0.1904) = 0.98096.$

1.5.9 $P(\text{no broken bulbs}) = \frac{83}{100} \times \frac{82}{99} \times \frac{81}{98} = 0.5682$

$P(\text{one broken bulb}) = P(\text{broken, not broken, not broken})$

$+ P(\text{not broken, broken, not broken}) + P(\text{not broken, not broken, broken})$

$= \left(\frac{17}{100} \times \frac{83}{99} \times \frac{82}{98} \right) + \left(\frac{83}{100} \times \frac{17}{99} \times \frac{82}{98} \right) + \left(\frac{83}{100} \times \frac{82}{99} \times \frac{17}{98} \right) = 0.3578$

$P(\text{no more than one broken bulb in the sample})$

$= P(\text{no broken bulbs}) + P(\text{one broken bulb})$

$= 0.5682 + 0.3578 = 0.9260$

1.5.11 $P(\text{drawing 2 green balls})$

$= P(\text{1st ball is green}) \times P(\text{2nd ball is green} \mid \text{1st ball is green})$

$= \frac{72}{169} \times \frac{71}{168} = 0.180$

$P(\text{two balls same color})$

$= P(\text{two red balls}) + P(\text{two blue balls}) + P(\text{two green balls})$

$= \left(\frac{43}{169} \times \frac{42}{168} \right) + \left(\frac{54}{169} \times \frac{53}{168} \right) + \left(\frac{72}{169} \times \frac{71}{168} \right) = 0.344$

$P(\text{two balls different colors}) = 1 - P(\text{two balls same color})$

$= 1 - 0.344 = 0.656$

1.5.13 $P(\text{same result on both throws}) = P(\text{both heads}) + P(\text{both tails})$

$= p^2 + (1 - p)^2 = 2p^2 - 2p + 1 = 2(p - 0.5)^2 + 0.5$

which is minimized when $p = 0.5$ (a fair coin).

1.5.15   (a) $\left(\frac{1}{2} \right)^5 = \frac{1}{32}$

(b) $1 \times \frac{5}{6} \times \frac{4}{6} = \frac{5}{9}$

(c) $P(BBR) + P(BRB) + P(RBB)$

$= \left(\frac{1}{2} \times \frac{1}{2} \times \frac{1}{2}\right) + \left(\frac{1}{2} \times \frac{1}{2} \times \frac{1}{2}\right) + \left(\frac{1}{2} \times \frac{1}{2} \times \frac{1}{2}\right)$

$= \frac{3}{8}$

(d) $P(BBR) + P(BRB) + P(RBB)$

$= \left(\frac{26}{52} \times \frac{25}{51} \times \frac{26}{50}\right) + \left(\frac{26}{52} \times \frac{26}{51} \times \frac{25}{50}\right) + \left(\frac{26}{52} \times \frac{26}{51} \times \frac{25}{50}\right)$

$= \frac{13}{34}$

1.5.17   Claims from clients in the same geographical area would not be independent of each other since they would all be affected by the same flooding events.

1.5.19 $(0.26 + 0.36 + 0.11) \times (0.26 + 0.36 + 0.11) = 0.5329$

1.6   Posterior Probabilities

1.6.1   (a)   The following information is given:

$P(\text{disease}) = 0.01$

$P(\text{no disease}) = 0.99$

$P(\text{positive blood test} \mid \text{disease}) = 0.97$

$P(\text{positive blood test} \mid \text{no disease}) = 0.06$

Therefore,

$P(\text{positive blood test}) = (P(\text{positive blood test} \mid \text{disease}) \times P(\text{disease}))$

$+ \ (P(\text{positive blood test} \mid \text{no disease}) \times P(\text{no disease}))$

$= (0.97 \times 0.01) + (0.06 \times 0.99) = 0.0691.$

(b) $P(\text{disease} \mid \text{positive blood test})$

$= \dfrac{P(\text{positive blood test} \cap \text{disease})}{P(\text{positive blood test})}$

$= \dfrac{P(\text{positive blood test} \mid \text{disease}) \times P(\text{disease})}{P(\text{positive blood test})}$

$= \dfrac{0.97 \times 0.01}{0.0691} = 0.1404$

(c) $P(\text{no disease} \mid \text{negative blood test})$

$= \dfrac{P(\text{no disease} \cap \text{negative blood test})}{P(\text{negative blood test})}$

$= \dfrac{P(\text{negative blood test} \mid \text{no disease}) \times P(\text{no disease})}{1 - P(\text{positive blood test})}$

$= \dfrac{(1 - 0.06) \times 0.99}{(1 - 0.0691)} = 0.9997$

1.6.3   (a) $P(\text{Section I}) = \frac{55}{100}$

(b) $P(\text{grade is A})$

$= (P(\text{A} \mid \text{Section I}) \times P(\text{Section I})) + (P(\text{A} \mid \text{Section II}) \times P(\text{Section II}))$

$= \left(\frac{10}{55} \times \frac{55}{100} \right) + \left(\frac{11}{45} \times \frac{45}{100} \right) = \frac{21}{100}$

(c) $P(\text{A} \mid \text{Section I}) = \frac{10}{55}$

(d) $P(\text{Section I} \mid \text{A}) = \dfrac{P(\text{A} \cap \text{Section I})}{P(\text{A})}$

$$= \frac{P(\text{Section I}) \times P(\text{A} \mid \text{Section I})}{P(\text{A})}$$

$$= \frac{\frac{55}{100} \times \frac{10}{55}}{\frac{21}{100}} = \frac{10}{21}$$

1.6.5 (a) $P(\text{fail}) = (0.02 \times 0.77) + (0.10 \times 0.11) + (0.14 \times 0.07) + (0.25 \times 0.05)$
$= 0.0487$

$P(C \mid \text{fail}) = \frac{0.14 \times 0.07}{0.0487} = 0.2012$

$P(D \mid \text{fail}) = \frac{0.25 \times 0.05}{0.0487} = 0.2567$

The answer is $0.2012 + 0.2567 = 0.4579$.

(b) $P(A \mid \text{did not fail}) = \frac{P(A) \times P(\text{did not fail} \mid A)}{P(\text{did not fail})}$

$= \frac{0.77 \times (1 - 0.02)}{1 - 0.0487} = 0.7932$

1.6.7 (a) $P(C) = 0.12$
$P(M) = 0.55$
$P(W) = 0.20$
$P(H) = 0.13$
$P(L \mid C) = 0.003$
$P(L \mid M) = 0.009$
$P(L \mid W) = 0.014$
$P(L \mid H) = 0.018$
Therefore,
$P(H \mid L) = \frac{P(L|H)P(H)}{P(L|C)P(C) + P(L|M)P(M) + P(L|W)P(W) + P(L|H)P(H)}$

$= \frac{0.018 \times 0.13}{(0.003 \times 0.12) + (0.009 \times 0.55) + (0.014 \times 0.20) + (0.018 \times 0.13)}$

$= 0.224$

(b) $P(M \mid L') = \frac{P(L'|M)P(M)}{P(L'|C)P(C) + P(L'|M)P(M) + P(L'|W)P(W) + P(L'|H)P(H)}$

$= \frac{0.991 \times 0.55}{(0.997 \times 0.12) + (0.991 \times 0.55) + (0.986 \times 0.20) + (0.982 \times 0.13)}$

$= 0.551$

1.6.9 $(0.08 \times 0.03) + (0.19 \times 0.14) + (0.26 \times 0.60) + (0.36 \times 0.77) + (0.11 \times 0.99) = 0.5711$

$$\frac{(0.08 \times 0.03) + (0.19 \times 0.14)}{(0.08 \times 0.03) + (0.19 \times 0.14) + (0.26 \times 0.60) + (0.36 \times 0.77) + (0.11 \times 0.99)} = 0.0508$$

1.7 Counting Techniques

1.7.1   (a) $7! = 7 \times 6 \times 5 \times 4 \times 3 \times 2 \times 1 = 5040$

(b) $8! = 8 \times 7! = 40320$

(c) $4! = 4 \times 3 \times 2 \times 1 = 24$

(d) $13! = 13 \times 12 \times 11 \times \ldots \times 1 = 6{,}227{,}020{,}800$

1.7.3   (a) $C_2^6 = \frac{6!}{(6-2)! \times 2!} = \frac{6 \times 5}{2} = 15$

(b) $C_4^8 = \frac{8!}{(8-4)! \times 4!} = \frac{8 \times 7 \times 6 \times 5}{24} = 70$

(c) $C_2^5 = \frac{5!}{(5-2)! \times 2!} = \frac{5 \times 4}{2} = 10$

(d) $C_6^{14} = \frac{14!}{(14-6)! \times 6!} = 3003$

1.7.5   The number of experimental configurations is $3 \times 4 \times 2 = 24$.

1.7.7   The number of rankings that can be assigned to the top 5 competitors is
$P_5^{20} = \frac{20!}{15!} = 20 \times 19 \times 18 \times 17 \times 16 = 1{,}860{,}480$.

The number of ways in which the best 5 competitors can be chosen is
$C_5^{20} = \frac{20!}{15! \times 5!} = 15504$.

1.7.9 $C_k^{n-1} + C_{k-1}^{n-1} = \frac{(n-1)!}{k!(n-1-k)!} + \frac{(n-1)!}{(k-1)!(n-k)!} = \frac{n!}{k!(n-k)!} \left(\frac{n-k}{n} + \frac{k}{n} \right) = \frac{n!}{k!(n-k)!}$
$= C_k^n$

This relationship can be interpreted in the following manner.

C_k^n is the number of ways that k balls can be selected from n balls. Suppose that one ball is red while the remaining $n-1$ balls are blue. Either all k balls selected are blue or one of the selected balls is red.

C_k^{n-1} is the number of ways k blue balls can be selected while C_{k-1}^{n-1} is the number of ways of selecting the one red ball and $k-1$ blue balls.

1.7.11   There are $n!$ ways in which n objects can be arranged in a line. If the line is made into a circle and rotations of the circle are considered to be indistinguishable, then there are n arrangements of the line corresponding to each arrangement of the circle. Consequently, there are $\frac{n!}{n} = (n-1)!$ ways to order the objects in a circle.

1.7.13   Consider 5 blocks, one block being Andrea and Scott and the other four blocks being the other four people. At the cinema these 5 blocks can be arranged in 5! ways, and then Andrea and Scott can be arranged in two different ways within their block, so that the total number of seating arrangements is $2 \times 5! = 240$.

Similarly, the total number of seating arrangements at the dinner table is $2 \times 4! = 48$.

If Andrea refuses to sit next to Scott then the number of seating arrangements can be obtained by subtraction. The total number of seating arrangements at the cinema is $720 - 240 = 480$ and the total number of seating arrangements at the dinner table is $120 - 48 = 72$.

1.7.15   (a) Using the result provided in the previous problem the answer is $\frac{12!}{3! \times 4! \times 5!} = 27720$.

      (b) Suppose that the balls in part (a) are labelled from 1 to 12. Then the positions of the three red balls in the line (where the places in the line are labelled 1 to 12) can denote which balls in part (a) are placed in the first box, the positions of the four blue balls in the line can denote which balls in part (a) are placed in the second box, and the positions of the five green balls in the line can denote which balls in part (a) are placed in the third box. Thus, there is a one-to-one correspondence between the positioning of the colored balls in part (b) and the arrangements of the balls in part (a) so that the problems are identical.

1.7.17 $\frac{15!}{3! \times 3! \times 3! \times 3! \times 3!} = 168,168,000$

1.7.19   The ordering of the visits can be made in $10! = 3,628,800$ different ways.

The number of different ways the ten cities be split into two groups of five cities is $C_5^{10} = 252$.

1.7.21   (a) $\dfrac{\binom{39}{8}}{\binom{52}{8}} = \frac{39}{52} \times \frac{38}{51} \times \frac{37}{50} \times \frac{36}{49} \times \frac{35}{48} \times \frac{34}{47} \times \frac{33}{46} \times \frac{32}{45} = 0.082$

(b) $\dfrac{\binom{13}{2} \times \binom{13}{2} \times \binom{13}{2} \times \binom{13}{2}}{\binom{52}{8}} = 0.049$

1.7.23 $5 \times 5 \times 5 = 125$

Chapter 2

Random Variables

2.1   Discrete Random Variables

2.1.1   (a)   Since
$$0.08 + 0.11 + 0.27 + 0.33 + P(X = 4) = 1$$
it follows that
$$P(X = 4) = 0.21.$$

(c) $F(0) = 0.08$
$F(1) = 0.19$
$F(2) = 0.46$
$F(3) = 0.79$
$F(4) = 1.00$

2.1.3

x_i	1	2	3	4	5	6	8	9	10
p_i	$\frac{1}{36}$	$\frac{2}{36}$	$\frac{2}{36}$	$\frac{3}{36}$	$\frac{2}{36}$	$\frac{4}{36}$	$\frac{2}{36}$	$\frac{1}{36}$	$\frac{2}{36}$
$F(x_i)$	$\frac{1}{36}$	$\frac{3}{36}$	$\frac{5}{36}$	$\frac{8}{36}$	$\frac{10}{36}$	$\frac{14}{36}$	$\frac{16}{36}$	$\frac{17}{36}$	$\frac{19}{36}$

27

x_i	12	15	16	18	20	24	25	30	36
p_i	$\frac{4}{36}$	$\frac{2}{36}$	$\frac{1}{36}$	$\frac{2}{36}$	$\frac{2}{36}$	$\frac{2}{36}$	$\frac{1}{36}$	$\frac{2}{36}$	$\frac{1}{36}$
$F(x_i)$	$\frac{23}{36}$	$\frac{25}{36}$	$\frac{26}{36}$	$\frac{28}{36}$	$\frac{30}{36}$	$\frac{32}{36}$	$\frac{33}{36}$	$\frac{35}{36}$	1

2.1.5

x_i	-5	-4	-3	-2	-1	0	1	2	3	4	6	8	10	12
p_i	$\frac{1}{36}$	$\frac{1}{36}$	$\frac{2}{36}$	$\frac{2}{36}$	$\frac{3}{36}$	$\frac{3}{36}$	$\frac{2}{36}$	$\frac{5}{36}$	$\frac{1}{36}$	$\frac{4}{36}$	$\frac{3}{36}$	$\frac{3}{36}$	$\frac{3}{36}$	$\frac{3}{36}$
$F(x_i)$	$\frac{1}{36}$	$\frac{2}{36}$	$\frac{4}{36}$	$\frac{6}{36}$	$\frac{9}{36}$	$\frac{12}{36}$	$\frac{14}{36}$	$\frac{19}{36}$	$\frac{20}{36}$	$\frac{24}{36}$	$\frac{27}{36}$	$\frac{30}{36}$	$\frac{33}{36}$	1

2.1.7 (a)

x_i	0	1	2	3	4	6	8	12
p_i	0.061	0.013	0.195	0.067	0.298	0.124	0.102	0.140

(b)

x_i	0	1	2	3	4	6	8	12
$F(x_i)$	0.061	0.074	0.269	0.336	0.634	0.758	0.860	1.000

(c) The most likely value is 4.

$$P(\text{not shipped}) = P(X \leq 1) = 0.074$$

2.1.9

x_i	1	2	3	4
p_i	$\frac{2}{5}$	$\frac{3}{10}$	$\frac{1}{5}$	$\frac{1}{10}$
$F(x_i)$	$\frac{2}{5}$	$\frac{7}{10}$	$\frac{9}{10}$	1

2.1.11   (a) The state space is $\{3, 4, 5, 6\}$.

(b) $P(X = 3) = P(MMM) = \frac{3}{6} \times \frac{2}{5} \times \frac{1}{4} = \frac{1}{20}$

$P(X = 4) = P(MMTM) + P(MTMM) + P(TMMM) = \frac{3}{20}$

$P(X = 5) = P(MMTTM) + P(MTMTM) + P(TMMTM)$
$+ P(MTTMM) + P(TMTMM) + P(TTMMM) = \frac{6}{20}$

Finally,
$P(X = 6) = \frac{1}{2}$
since the probabilities sum to one, or since the final appointment made is equally likely to be on a Monday or on a Tuesday.

$P(X \leq 3) = \frac{1}{20}$
$P(X \leq 4) = \frac{4}{20}$
$P(X \leq 5) = \frac{10}{20}$
$P(X \leq 6) = 1$

2.2   Continuous Random Variables

2.2.1   (a) Continuous

(b) Discrete

(c) Continuous

(d) Continuous

(e) Discrete

(f) This depends on what level of accuracy to which it is measured. It could be considered to be either discrete or continuous.

2.2.3   (a) Since

$$\int_{-2}^{0} \left(\frac{15}{64} + \frac{x}{64} \right) dx = \frac{7}{16}$$

and

$$\int_{0}^{3} \left(\frac{3}{8} + cx \right) dx = \frac{9}{8} + \frac{9c}{2}$$

it follows that

$$\frac{7}{16} + \frac{9}{8} + \frac{9c}{2} = 1$$

which gives $c = -\frac{1}{8}$.

(b) $P(-1 \leq X \leq 1) = \int_{-1}^{0} \left(\frac{15}{64} + \frac{x}{64} \right) dx + \int_{0}^{1} \left(\frac{3}{8} - \frac{x}{8} \right) dx$

$= \frac{69}{128}$

(c) $F(x) = \int_{-2}^{x} \left(\frac{15}{64} + \frac{y}{64} \right) dy$

$= \frac{x^2}{128} + \frac{15x}{64} + \frac{7}{16}$

for $-2 \leq x \leq 0$

$F(x) = \frac{7}{16} + \int_{0}^{x} \left(\frac{3}{8} - \frac{y}{8} \right) dy$

$= -\frac{x^2}{16} + \frac{3x}{8} + \frac{7}{16}$

for $0 \leq x \leq 3$

2.2.5　(a) Since $F(\infty) = 1$ it follows that $A = 1$.
Then $F(0) = 0$ gives $1 + B = 0$ so that $B = -1$ and
$F(x) = 1 - e^{-x}$.

(b) $P(2 \leq X \leq 3) = F(3) - F(2)$
$= e^{-2} - e^{-3} = 0.0855$

(c) $f(x) = \frac{dF(x)}{dx} = e^{-x}$
for $x \geq 0$

2.2.7　(a) Since
$$F(0) = A + B\ln(2) = 0$$
and
$$F(10) = A + B\ln(32) = 1$$
it follows that $A = -0.25$ and $B = \frac{1}{\ln(16)} = 0.361$.

(b) $P(X > 2) = 1 - F(2) = 0.5$

(c) $f(x) = \frac{dF(x)}{dx} = \frac{1.08}{3x+2}$
for $0 \leq x \leq 10$

2.2.9　(a) Since $F(0) = 0$ and $F(50) = 1$
it follows that $A = 1.0007$ and $B = -125.09$.

(b) $P(X \leq 10) = F(10) = 0.964$

(c) $P(X \geq 30) = 1 - F(30) = 1 - 0.998 = 0.002$

(d) $f(r) = \frac{dF(r)}{dr} = \frac{375.3}{(r+5)^4}$
for $0 \leq r \leq 50$

2.2.11   (a)   Since

$$\int_{10}^{11} Ax(130 - x^2) \, dx = 1$$

it follows that

$$A = \tfrac{4}{819}.$$

(b) $F(x) = \int_{10}^{x} \frac{4y(130 - y^2)}{819} \, dy$

$$= \tfrac{4}{819} \left(65x^2 - \tfrac{x^4}{4} - 4000 \right)$$

for $10 \leq x \leq 11$

(c) $F(10.5) - F(10.25) = 0.623 - 0.340 = 0.283$

2.3 The Expectation of a Random Variable

2.3.1 $E(X) = (0 \times 0.08) + (1 \times 0.11) + (2 \times 0.27) + (3 \times 0.33) + (4 \times 0.21)$
$= 2.48$

2.3.3 With replacement:
$E(X) = (0 \times 0.5625) + (1 \times 0.3750) + (2 \times 0.0625)$
$= 0.5$

Without replacement:
$E(X) = (0 \times 0.5588) + (1 \times 0.3824) + (2 \times 0.0588)$
$= 0.5$

2.3.5

x_i	2	3	4	5	6	7	8	9	10	15
p_i	$\frac{1}{13}$	$\frac{1}{13}$	$\frac{1}{13}$	$\frac{1}{13}$	$\frac{1}{13}$	$\frac{1}{13}$	$\frac{1}{13}$	$\frac{1}{13}$	$\frac{1}{13}$	$\frac{4}{13}$

$$E(X) = \left(2 \times \tfrac{1}{13}\right) + \left(3 \times \tfrac{1}{13}\right) + \left(4 \times \tfrac{1}{13}\right) + \left(5 \times \tfrac{1}{13}\right) + \left(6 \times \tfrac{1}{13}\right)$$
$$+ \left(7 \times \tfrac{1}{13}\right) + \left(8 \times \tfrac{1}{13}\right) + \left(9 \times \tfrac{1}{13}\right) + \left(10 \times \tfrac{1}{13}\right) + \left(15 \times \tfrac{4}{13}\right)$$
$$= \$8.77$$

If $9 is paid to play the game, the expected loss would be 23 cents.

2.3.7 $P(\text{three sixes are rolled}) = \frac{1}{6} \times \frac{1}{6} \times \frac{1}{6}$
$= \frac{1}{216}$
so that
$E(\text{net winnings}) = \left(-\$1 \ \times \ \tfrac{215}{216}\right) + \left(\$499 \ \times \ \tfrac{1}{216}\right)$
$= \$1.31.$

If you can play the game a large number of times then you should play the game as often as you can.

2.3.9

x_i	0	1	2	3	4	5
p_i	0.1680	0.2816	0.2304	0.1664	0.1024	0.0512

$E(\text{payment}) = (0 \times 0.1680) + (1 \times 0.2816) + (2 \times 0.2304)$

$+ (3 \times 0.1664) + (4 \times 0.1024) + (5 \times 0.0512)$

$= 1.9072$

$E(\text{winnings}) = \$2 - \$1.91 = \$0.09$

The expected winnings increase to 9 cents per game.

Increasing the probability of scoring a three reduces the expected value of the difference in the scores of the two dice.

2.3.11   (a) $E(X) = \int_0^4 x \, \frac{x}{8} \, dx = 2.67$

(b) Solving $F(x) = 0.5$ gives $x = \sqrt{8} = 2.83$.

2.3.13 $E(X) = \int_0^{10} \frac{\theta}{e^{10}-11} \left(e^{10-\theta} - 1\right) d\theta = 0.9977$

Solving $F(\theta) = 0.5$ gives $\theta = 0.6927$.

2.3.15   Let $f(x)$ be a probability density function that is symmetric about the point μ, so that $f(\mu + x) = f(\mu - x)$.

Then

$E(X) = \int_{-\infty}^{\infty} x f(x) \, dx$

which under the transformation $x = \mu + y$ gives

$$E(X) = \int_{-\infty}^{\infty} (\mu + y) f(\mu + y) \, dy$$

$$= \mu \int_{-\infty}^{\infty} f(\mu + y) \, dy \; + \; \int_{0}^{\infty} y \left(f(\mu + y) - f(\mu - y) \right) dy$$

$$= (\mu \times 1) + 0 = \mu.$$

2.3.17   (a) $E(X) = \int_{10}^{11} \frac{4x^2(130 - x^2)}{819} \, dx$

            $= 10.418234$

     (b)   Solving $F(x) = 0.5$ gives the median as 10.385.

2.3.19 $(0 \times 0.38) + (1 \times 0.44) + (2 \times 0.15) + (3 \times 0.03) = 0.83$

2.4   The Variance of a Random Variable

2.4.1   (a) $E(X) = \left(-2 \times \frac{1}{3}\right) + \left(1 \times \frac{1}{6}\right) + \left(4 \times \frac{1}{3}\right) + \left(6 \times \frac{1}{6}\right)$

$\qquad = \frac{11}{6}$

(b) $\mathrm{Var}(X) = \left(\frac{1}{3} \times \left(-2 - \frac{11}{6}\right)^2\right) + \left(\frac{1}{6} \times \left(1 - \frac{11}{6}\right)^2\right)$

$\qquad + \left(\frac{1}{3} \times \left(4 - \frac{11}{6}\right)^2\right) + \left(\frac{1}{6} \times \left(6 - \frac{11}{6}\right)^2\right)$

$\qquad = \frac{341}{36}$

(c) $E(X^2) = \left(\frac{1}{3} \times (-2)^2\right) + \left(\frac{1}{6} \times 1^2\right) + \left(\frac{1}{3} \times 4^2\right) + \left(\frac{1}{6} \times 6^2\right)$

$\qquad = \frac{77}{6}$

$\qquad \mathrm{Var}(X) = E(X^2) - E(X)^2 = \frac{77}{6} - \left(\frac{11}{6}\right)^2 = \frac{341}{36}$

2.4.3 $E(X^2) = \left(1^2 \times \frac{2}{5}\right) + \left(2^2 \times \frac{3}{10}\right) + \left(3^2 \times \frac{1}{5}\right) + \left(4^2 \times \frac{1}{10}\right)$

$\qquad = 5$

Then $E(X) = 2$ so that

$\mathrm{Var}(X) = 5 - 2^2 = 1$

and $\sigma = 1$.

2.4.5   (a) $E(X^2) = \int_4^6 x^2 \frac{1}{x \ln(1.5)} \, dx = 24.6630$

Then $E(X) = 4.9326$ so that

$\mathrm{Var}(X) = 24.6630 - 4.9326^2 = 0.3325$.

(b) $\sigma = \sqrt{0.3325} = 0.5766$

(c) Solving $F(x) = 0.25$ gives $x = 4.43$.

Solving $F(x) = 0.75$ gives $x = 5.42$.

    (d) The interquartile range is $5.42 - 4.43 = 0.99$.

2.4.7   (a) $E(X^2) = \int_{0.125}^{0.5} x^2 \, 5.5054 \, (0.5 - (x - 0.25)^2) \, dx = 0.1073$

       Then $E(X) = 0.3095$ so that
       $\mathrm{Var}(X) = 0.1073 - 0.3095^2 = 0.0115$.

    (b) $\sigma = \sqrt{0.0115} = 0.107$

    (c) Solving $F(x) = 0.25$ gives $x = 0.217$.
       Solving $F(x) = 0.75$ gives $x = 0.401$.

    (d) The interquartile range is $0.401 - 0.217 = 0.184$.

2.4.9   (a) $E(X^2) = \int_0^{50} \frac{375.3 \, r^2}{(r+5)^4} \, dr = 18.80$

       Then $E(X) = 2.44$ so that
       $\mathrm{Var}(X) = 18.80 - 2.44^2 = 12.8$.

    (b) $\sigma = \sqrt{12.8} = 3.58$

    (c) Solving $F(r) = 0.25$ gives $r = 0.50$.
       Solving $F(r) = 0.75$ gives $r = 2.93$.

    (d) The interquartile range is $2.93 - 0.50 = 2.43$.

2.4.11   The interval $(109.55, 112.05)$ is $(\mu - 2.5c, \mu + 2.5c)$
       so Chebyshev's inequality gives:
       $P(109.55 \leq X \leq 112.05) \geq 1 - \frac{1}{2.5^2} = 0.84$

2.4.13   (a) $E(X^2) = \int_{10}^{11} \frac{4x^3(130 - x^2)}{819} \, dx$

       $= 108.61538$

Therefore,

$\text{Var}(X) = E(X^2) - (E(X))^2 = 108.61538 - 10.418234^2 = 0.0758$

and the standard deviation is $\sqrt{0.0758} = 0.275$.

(b) Solving $F(x) = 0.8$ gives the 80th percentile of the resistance as 10.69,

and solving $F(x) = 0.1$ gives the 10th percentile of the resistance as 10.07.

2.4.15 $E(X) = (-1 \times 0.25) + (1 \times 0.4) + (4 \times 0.35)$

$= \$1.55$

$E(X^2) = ((-1)^2 \times 0.25) + (1^2 \times 0.4) + (4^2 \times 0.35)$

$= 6.25$

Therefore, the variance is

$E(X^2) - (E(X))^2 = 6.25 - 1.55^2 = 3.8475$

and the standard deviation is $\sqrt{3.8475} = \$1.96$.

2.4.17   (a) $E(X) = (2 \times 0.11) + (3 \times 0.19) + (4 \times 0.55) + (5 \times 0.15)$

$= 3.74$

(b) $E(X^2) = (2^2 \times 0.11) + (3^2 \times 0.19) + (4^2 \times 0.55) + (5^2 \times 0.15)$

$= 14.70$

Therefore,

$\text{Var}(X) = 14.70 - 3.74^2 = 0.7124$

and the standard deviation is $\sqrt{0.7124} = 0.844$.

2.5 Jointly Distributed Random Variables

2.5.1   (a) $P(0.8 \leq X \leq 1, 25 \leq Y \leq 30)$

$$= \int_{x=0.8}^{1} \int_{y=25}^{30} \left(\frac{39}{400} - \frac{17(x-1)^2}{50} - \frac{(y-25)^2}{10000} \right) dx \, dy$$

$$= 0.092$$

(b) $E(Y) = \int_{20}^{35} y \left(\frac{83}{1200} - \frac{(y-25)^2}{10000} \right) dy = 27.36$

$E(Y^2) = \int_{20}^{35} y^2 \left(\frac{83}{1200} - \frac{(y-25)^2}{10000} \right) dy = 766.84$

$\text{Var}(Y) = E(Y^2) - E(Y)^2 = 766.84 - (27.36)^2 = 18.27$

$\sigma_Y = \sqrt{18.274} = 4.27$

(c) $E(Y|X = 0.55) = \int_{20}^{35} y \left(0.073 - \frac{(y-25)^2}{3922.5} \right) dy = 27.14$

$E(Y^2|X = 0.55) = \int_{20}^{35} y^2 \left(0.073 - \frac{(y-25)^2}{3922.5} \right) dy = 753.74$

$\text{Var}(Y|X = 0.55) = E(Y^2|X = 0.55) - E(Y|X = 0.55)^2$

$$= 753.74 - (27.14)^2 = 17.16$$

$\sigma_{Y|X=0.55} = \sqrt{17.16} = 4.14$

2.5.3   (a) Since

$$\int_{x=-2}^{3} \int_{y=4}^{6} A(x-3)y \, dx \, dy = 1$$

it follows that $A = -\frac{1}{125}$.

(b) $P(0 \leq X \leq 1, 4 \leq Y \leq 5)$

$$= \int_{x=0}^{1} \int_{y=4}^{5} \frac{(3-x)y}{125} \, dx \, dy$$

$$= \frac{9}{100}$$

(c) $f_X(x) = \int_4^6 \frac{(3-x)y}{125}\ dy = \frac{2(3-x)}{25}$

for $-2 \le x \le 3$

$f_Y(y) = \int_{-2}^3 \frac{(3-x)y}{125}\ dx = \frac{y}{10}$

for $4 \le x \le 6$

(d) The random variables X and Y are independent since

$f_X(x) \times f_Y(y) = f(x,y)$

and the ranges of the random variables are not related.

(e) Since the random variables are independent it follows that

$f_{X|Y=5}(x)$ is equal to $f_X(x)$.

2.5.5   (a) Since

$\int_{x=1}^2 \int_{y=0}^3 A(e^{x+y}\ +\ e^{2x-y})\ dx\ dy = 1$

it follows that $A = 0.00896$.

(b) $P(1.5 \le X \le 2, 1 \le Y \le 2)$

$= \int_{x=1.5}^2 \int_{y=1}^2 0.00896\ (e^{x+y}\ +\ e^{2x-y})\ dx\ dy$

$= 0.158$

(c) $f_X(x) = \int_0^3 0.00896\ (e^{x+y}\ +\ e^{2x-y})\ dy$

$= 0.00896\ (e^{x+3} - e^{2x-3} - e^x + e^{2x})$

for $1 \le x \le 2$

$f_Y(y) = \int_1^2 0.00896\ (e^{x+y}\ +\ e^{2x-y})\ dx$

$= 0.00896\ (e^{2+y} + 0.5e^{4-y} - e^{1+y} - 0.5e^{2-y})$

for $0 \le y \le 3$

(d) No, since $f_X(x) \times f_Y(y) \ne f(x,y)$.

(e) $f_{X|Y=0}(x) = \frac{f(x,0)}{f_Y(0)} = \frac{e^x + e^{2x}}{28.28}$

2.5.7   (a)

X\Y	0	1	2	p_{i+}
0	$\frac{4}{16}$	$\frac{4}{16}$	$\frac{1}{16}$	$\frac{9}{16}$
1	$\frac{4}{16}$	$\frac{2}{16}$	0	$\frac{6}{16}$
2	$\frac{1}{16}$	0	0	$\frac{1}{16}$
p_{+j}	$\frac{9}{16}$	$\frac{6}{16}$	$\frac{1}{16}$	1

(b) See the table above.

(c) No, the random variables X and Y are not independent.
    For example,
    $p_{22} \neq p_{2+} \times p_{+2}$.

(d) $E(X) = \left(0 \times \frac{9}{16}\right) + \left(1 \times \frac{6}{16}\right) + \left(2 \times \frac{1}{16}\right) = \frac{1}{2}$

   $E(X^2) = \left(0^2 \times \frac{9}{16}\right) + \left(1^2 \times \frac{6}{16}\right) + \left(2^2 \times \frac{1}{16}\right) = \frac{5}{8}$

   $\text{Var}(X) = E(X^2) - E(X)^2 = \frac{5}{8} - \left(\frac{1}{2}\right)^2 = \frac{3}{8} = 0.375$

   The random variable Y has the same mean and variance as X.

(e) $E(XY) = 1 \times 1 \times p_{11} = \frac{1}{8}$

   $\text{Cov}(X,Y) = E(XY) - (E(X) \times E(Y))$

   $= \frac{1}{8} - \left(\frac{1}{2} \times \frac{1}{2}\right) = -\frac{1}{8}$

(f) $\text{Corr}(X,Y) = \frac{\text{Cov}(X,Y)}{\sqrt{\text{Var}(X)\text{Var}(Y)}} = -\frac{1}{3}$

(g) $P(Y = 0 | X = 0) = \frac{p_{00}}{p_{0+}} = \frac{4}{9}$

$P(Y = 1 | X = 0) = \frac{p_{01}}{p_{0+}} = \frac{4}{9}$

$P(Y = 2 | X = 0) = \frac{p_{02}}{p_{0+}} = \frac{1}{9}$

$P(Y = 0 | X = 1) = \frac{p_{10}}{p_{1+}} = \frac{2}{3}$

$P(Y = 1 | X = 1) = \frac{p_{11}}{p_{1+}} = \frac{1}{3}$

$P(Y = 2 | X = 1) = \frac{p_{12}}{p_{1+}} = 0$

2.5.9   (a) $P(\text{same score}) = P(X = 1, Y = 1) + P(X = 2, Y = 2)$

$+ P(X = 3, Y = 3) + P(X = 4, Y = 4)$

$= 0.80$

(b) $P(X < Y) = P(X = 1, Y = 2) + P(X = 1, Y = 3)$

$+ P(X = 1, Y = 4) + P(X = 2, Y = 3)$

$+ P(X = 2, Y = 4) + P(X = 3, Y = 4)$

$= 0.07$

(c)

x_i	1	2	3	4
p_{i+}	0.12	0.20	0.30	0.38

$E(X) = (1 \times 0.12) + (2 \times 0.20) + (3 \times 0.30) + (4 \times 0.38) = 2.94$

$E(X^2) = (1^2 \times 0.12) + (2^2 \times 0.20) + (3^2 \times 0.30) + (4^2 \times 0.38) = 9.70$

$\text{Var}(X) = E(X^2) - E(X)^2 = 9.70 - (2.94)^2 = 1.0564$

(d)

y_j	1	2	3	4
p_{+j}	0.14	0.21	0.30	0.35

$$E(Y) = (1 \times 0.14) + (2 \times 0.21) + (3 \times 0.30) + (4 \times 0.35) = 2.86$$

$$E(Y^2) = (1^2 \times 0.14) + (2^2 \times 0.21) + (3^2 \times 0.30) + (4^2 \times 0.35) = 9.28$$

$$\text{Var}(Y) = E(Y^2) - E(Y)^2 = 9.28 - (2.86)^2 = 1.1004$$

(e) The scores are not independent.

For example, $p_{11} \neq p_{1+} \times p_{+1}$.

The scores would not be expected to be independent since they apply to the two inspectors' assessments of the same building. If they were independent it would suggest that one of the inspectors is randomly assigning a safety score without paying any attention to the actual state of the building.

(f) $P(Y = 1 | X = 3) = \frac{p_{31}}{p_{3+}} = \frac{1}{30}$

$P(Y = 2 | X = 3) = \frac{p_{32}}{p_{3+}} = \frac{3}{30}$

$P(Y = 3 | X = 3) = \frac{p_{33}}{p_{3+}} = \frac{24}{30}$

$P(Y = 4 | X = 3) = \frac{p_{34}}{p_{3+}} = \frac{2}{30}$

(g) $E(XY) = \sum_{i=1}^{4} \sum_{j=1}^{4} i \, j \, p_{ij} = 9.29$

$\text{Cov}(X, Y) = E(XY) - (E(X) \times E(Y))$
$= 9.29 - (2.94 \times 2.86) = 0.8816$

(h) $\text{Corr}(X, Y) = \frac{\text{Cov}(X,Y)}{\sqrt{\text{Var}X \ \text{Var}Y}} = \frac{0.8816}{\sqrt{1.0564 \times 1.1004}} = 0.82$

A high positive correlation indicates that the inspectors are consistent.

The closer the correlation is to one the more consistent the inspectors are.

2.6 Combinations and Functions of Random variables

2.6.1   (a) $E(3X + 7) = 3E(X) + 7 = 13$

$\text{Var}(3X + 7) = 3^2\text{Var}(X) = 36$

(b) $E(5X - 9) = 5E(X) - 9 = 1$

$\text{Var}(5X - 9) = 5^2\text{Var}(X) = 100$

(c) $E(2X + 6Y) = 2E(X) + 6E(Y) = -14$

$\text{Var}(2X + 6Y) = 2^2\text{Var}(X) + 6^2\text{Var}(Y) = 88$

(d) $E(4X - 3Y) = 4E(X) - 3E(Y) = 17$

$\text{Var}(4X - 3Y) = 4^2\text{Var}(X) + 3^2\text{Var}(Y) = 82$

(e) $E(5X - 9Z + 8) = 5E(X) - 9E(Z) + 8 = -54$

$\text{Var}(5X - 9Z + 8) = 5^2\text{Var}(X) + 9^2\text{Var}(Z) = 667$

(f) $E(-3Y - Z - 5) = -3E(Y) - E(Z) - 5 = -4$

$\text{Var}(-3Y - Z - 5) = (-3)^2\text{Var}(Y) + (-1)^2\text{Var}(Z) = 25$

(g) $E(X + 2Y + 3Z) = E(X) + 2E(Y) + 3E(Z) = 20$

$\text{Var}(X + 2Y + 3Z) = \text{Var}(X) + 2^2\text{Var}(Y) + 3^2\text{Var}(Z) = 75$

(h) $E(6X + 2Y - Z + 16) = 6E(X) + 2E(Y) - E(Z) + 16 = 14$

$\text{Var}(6X + 2Y - Z + 16) = 6^2\text{Var}(X) + 2^2\text{Var}(Y) + (-1)^2\text{Var}(Z) = 159$

2.6.3 $E(Y) = 3E(X_1) = 3\mu$

$\text{Var}(Y) = 3^2\text{Var}(X_1) = 9\sigma^2$

$E(Z) = E(X_1) + E(X_2) + E(X_3) = 3\mu$

$\text{Var}(Z) = \text{Var}(X_1) + \text{Var}(X_2) + \text{Var}(X_3) = 3\sigma^2$

The random variables Y and Z have the same mean
but Z has a smaller variance than Y.

2.6.5   Let the random variable X_i be the winnings from the ith game.
Then

$$E(X_i) = \left(10 \times \tfrac{1}{8}\right) + \left((-1) \times \tfrac{7}{8}\right) = \tfrac{3}{8}$$

and

$$E(X_i^2) = \left(10^2 \times \tfrac{1}{8}\right) + \left((-1)^2 \times \tfrac{7}{8}\right) = \tfrac{107}{8}$$

so that

$$\text{Var}(X_i) = E(X_i^2) - (E(X_i))^2 = \tfrac{847}{64}.$$

The total winnings from 50 (independent) games is

$$Y = X_1 + \ldots + X_{50}$$

and

$$E(Y) = E(X_1) + \ldots + E(X_{50}) = 50 \times \tfrac{3}{8} = \tfrac{75}{4} = \$18.75$$

with

$$\text{Var}(Y) = \text{Var}(X_1) + \ldots + \text{Var}(X_{50}) = 50 \times \tfrac{847}{64} = 661.72$$

so that $\sigma_Y = \sqrt{661.72} = \$25.72$.

2.6.7   Let the random variable X_i be equal to 1 if an ace is drawn on the ith
drawing (which happens with a probability of $\tfrac{1}{13}$) and equal to 0 if an
ace is not drawn on the ith drawing (which happens with a probability of
$\tfrac{12}{13}$).

Then the total number of aces drawn is $Y = X_1 + \ldots + X_{10}$.

Notice that $E(X_i) = \tfrac{1}{13}$ so that regardless of whether the drawing is
performed with or without replacement it follows that

$$E(Y) = E(X_1) + \ldots + E(X_{10}) = \tfrac{10}{13}.$$

Also, notice that $E(X_i^2) = \frac{1}{13}$ so that

$\text{Var}(X_i) = \frac{1}{13} - \left(\frac{1}{13}\right)^2 = \frac{12}{169}$.

If the drawings are made *with replacement* then the random variables X_i are independent so that

$\text{Var}(Y) = \text{Var}(X_1) + \ldots + \text{Var}(X_{10}) = \frac{120}{169}$.

However, if the drawings are made *without replacement* then the random variables X_i are not independent.

2.6.9   (a) Since

$$\int_0^2 A(1 - (r-1)^2) \, dr = 1$$

it follows that $A = \frac{3}{4}$.

This gives

$$F_R(r) = \frac{3r^2}{4} - \frac{r^3}{4}$$

for $0 \le r \le 2$.

(b) $V = \frac{4}{3}\pi r^3$

Since

$$F_V(v) = P(V \le v) = P\left(\frac{4}{3}\pi r^3 \le v\right) = F_R\left(\left(\frac{3v}{4\pi}\right)^{1/3}\right)$$

it follows that

$$f_V(v) = \frac{1}{2}\left(\frac{3}{4\pi}\right)^{2/3} v^{-1/3} - \frac{3}{16\pi}$$

for $0 \le v \le \frac{32\pi}{3}$.

(c) $E(V) = \int_0^{\frac{32\pi}{3}} v \, f_V(v) \, dv = \frac{32\pi}{15}$

2.6.11   (a) The return has an expectation of \$100,
a standard deviation of \$20,
and a variance of 400.

(b) The return has an expectation of $100,

a standard deviation of $30,

and a variance of 900.

(c) The return from fund A has an expectation of $50,

a standard deviation of $10,

and a variance of 100.

The return from fund B has an expectation of $50,

a standard deviation of $15,

and a variance of 225.

Therefore, the total return has an expectation of $100 and a variance of 325, so that the standard deviation is $18.03.

(d) The return from fund A has an expectation of $0.1x,

a standard deviation of $0.02x,

and a variance of $0.0004x^2$.

The return from fund B has an expectation of $0.1(1000 - x)$,

a standard deviation of $0.03(1000 - x)$,

and a variance of $0.0009(1000 - x)^2$.

Therefore, the total return has an expectation of $100

and a variance of $0.0004x^2 + 0.0009(1000 - x)^2$.

This variance is minimized by taking $x = \$692$,

and the minimum value of the variance is 276.9

which corresponds to a standard deviation of $16.64.

This problem illustrates that the variability of the return on an investment can be reduced by *diversifying* the investment, so that it is spread over several funds.

2.6.13   (a) The mean is

$$E(X) = \left(\tfrac{1}{3} \times E(X_1)\right) + \left(\tfrac{1}{3} \times E(X_2)\right) + \left(\tfrac{1}{3} \times E(X_3)\right)$$

$$= \left(\tfrac{1}{3} \times 59\right) + \left(\tfrac{1}{3} \times 67\right) + \left(\tfrac{1}{3} \times 72\right) = 66$$

The variance is

$$\text{Var}(X) = \left(\left(\tfrac{1}{3}\right)^2 \times \text{Var}(X_1)\right) + \left(\left(\tfrac{1}{3}\right)^2 \times \text{Var}(X_2)\right) + \left(\left(\tfrac{1}{3}\right)^2 \times \text{Var}(X_3)\right)$$

$$= \left(\left(\tfrac{1}{3}\right)^2 \times 10^2\right) + \left(\left(\tfrac{1}{3}\right)^2 \times 13^2\right) + \left(\left(\tfrac{1}{3}\right)^2 \times 4^2\right) = \tfrac{95}{3}$$

so that the standard deviation is $\sqrt{95/3} = 5.63$.

(b) The mean is

$$E(X) = (0.4 \times E(X_1)) + (0.4 \times E(X_2)) + (0.2 \times E(X_3))$$

$$= (0.4 \times 59) + (0.4 \times 67) + (0.2 \times 72) = 64.8.$$

The variance is

$$\text{Var}(X) = (0.4^2 \times \text{Var}(X_1)) + (0.4^2 \times \text{Var}(X_2)) + (0.2^2 \times \text{Var}(X_3))$$

$$= (0.4^2 \times 10^2) + (0.4^2 \times 13^2) + (0.2^2 \times 4^2) = 43.68$$

so that the standard deviation is $\sqrt{43.68} = 6.61$.

2.6.15 (a) The mean is $\mu = 65.90$.
The standard deviation is $\frac{\sigma}{\sqrt{5}} = \frac{0.32}{\sqrt{5}} = 0.143$.

(b) The mean is $8\mu = 8 \times 65.90 = 527.2$.
The standard deviation is $\sqrt{8}\sigma = \sqrt{8} \times 0.32 = 0.905$.

2.6.17 When a die is rolled once the expectation is 3.5 and the standard deviation is 1.71 (see Games of Chance in section 2.4).

Therefore, the sum of eighty die rolls has an expectation of $80 \times 3.5 = 280$ and a standard deviation of $\sqrt{80} \times 1.71 = 15.3$.

2.6.19 Let X be the temperature in Fahrenheit and let Y be the temperature in Centigrade.

$$E(Y) = E\left(\tfrac{5(X-32)}{9}\right) = \left(\tfrac{5(E(X)-32)}{9}\right) = \left(\tfrac{5(110-32)}{9}\right) = 43.33$$

$$\text{Var}(Y) = \text{Var}\left(\tfrac{5(X-32)}{9}\right) = \left(\tfrac{5^2\text{Var}(X)}{9^2}\right) = \left(\tfrac{5^2 \times 2.2^2}{9^2}\right) = 1.49$$

The standard deviation is $\sqrt{1.49} = 1.22$.

2.6.21   The inequality $\frac{56}{\sqrt{n}} \le 10$ is satisfied for $n \ge 32$.

2.6.23   (a) $\sqrt{14000^2 + 14000^2} = \$19,799$

          (b) $14000/\sqrt{2} = \$9,899$

Chapter 3

Discrete Probability Distributions

3.1 The Binomial Distribution

3.1.1   (a) $P(X = 3) = \binom{10}{3} \times 0.12^3 \times 0.88^7 = 0.0847$

      (b) $P(X = 6) = \binom{10}{6} \times 0.12^6 \times 0.88^4 = 0.0004$

      (c) $P(X \leq 2) = P(X = 0) + P(X = 1) + P(X = 2)$
$$= 0.2785 + 0.3798 + 0.2330$$
$$= 0.8913$$

      (d) $P(X \geq 7) = P(X = 7) + P(X = 8) + P(X = 9) + P(X = 10)$
$$= 3.085 \times 10^{-5}$$

      (e) $E(X) = 10 \times 0.12 = 1.2$

      (f) $\mathrm{Var}(X) = 10 \times 0.12 \times 0.88 = 1.056$

3.1.3 $X \sim B(6, 0.5)$

50

x_i	0	1	2	3	4	5	6
p_i	0.0156	0.0937	0.2344	0.3125	0.2344	0.0937	0.0156

$E(X) = 6 \times 0.5 = 3$

$\text{Var}(X) = 6 \times 0.5 \times 0.5 = 1.5$

$\sigma = \sqrt{1.5} = 1.22$

$X \sim B(6, 0.7)$

x_i	0	1	2	3	4	5	6
p_i	0.0007	0.0102	0.0595	0.1852	0.3241	0.3025	0.1176

$E(X) = 6 \times 0.7 = 4.2$

$\text{Var}(X) = 6 \times 0.7 \times 0.3 = 1.26$

$\sigma = \sqrt{1.5} = 1.12$

3.1.5   (a) $P\left(B\left(8, \frac{1}{2}\right) = 5\right) = 0.2187$

       (b) $P\left(B\left(8, \frac{1}{6}\right) = 1\right) = 0.3721$

       (c) $P\left(B\left(8, \frac{1}{6}\right) = 0\right) = 0.2326$

3.1.7   Let the random variable X be the number of employees taking sick leave. Then $X \sim B(180, 0.35)$.

Therefore, the *proportion* of the workforce who need to take sick leave is

$$Y = \frac{X}{180}$$

so that

$$E(Y) = \frac{E(X)}{180} = \frac{180 \times 0.35}{180} = 0.35$$

and

$$\text{Var}(Y) = \frac{\text{Var}(X)}{180^2} = \frac{180 \times 0.35 \times 0.65}{180^2} = 0.0013.$$

In general, the variance is

$$\text{Var}(Y) = \frac{\text{Var}(X)}{180^2} = \frac{180 \times p \times (1-p)}{180^2} = \frac{p \times (1-p)}{180}$$

which is maximized when $p = 0.5$.

3.1.9 $X \sim B(18, 0.6)$

(a) $P(X = 8) + P(X = 9) + P(X = 10)$

$$= \binom{18}{8} \times 0.6^8 \times 0.4^{10} + \binom{18}{9} \times 0.6^9 \times 0.4^9 + \binom{18}{10} \times 0.6^{10} \times 0.4^8$$

$$= 0.0771 + 0.1284 + 0.1734 = 0.3789$$

(b) $P(X = 0) + P(X = 1) + P(X = 2) + P(X = 3) + P(X = 4)$

$$= \binom{18}{0} \times 0.6^0 \times 0.4^{18} + \binom{18}{1} \times 0.6^1 \times 0.4^{17} + \binom{18}{2} \times 0.6^2 \times 0.4^{16}$$

$$+ \binom{18}{3} \times 0.6^3 \times 0.4^{15} + \binom{18}{4} \times 0.6^4 \times 0.4^{14}$$

$$= 0.0013$$

3.1.11 $P(B(10, 0.65) \geq 5) = 0.905$

3.2 The Geometric and Negative Binomial Distributions

3.2.1   (a) $P(X = 4) = (1 - 0.7)^3 \times 0.7 = 0.0189$

       (b) $P(X = 1) = (1 - 0.7)^0 \times 0.7 = 0.7$

       (c) $P(X \leq 5) = 1 - (1 - 0.7)^5 = 0.9976$

       (d) $P(X \geq 8) = 1 - P(X \leq 7) = (1 - 0.7)^7 = 0.0002$

3.2.5   (a)   Consider a geometric distribution with parameter $p = 0.09$.
          $(1 - 0.09)^3 \times 0.09 = 0.0678$

       (b)   Consider a negative binomial distribution with parameters $p = 0.09$ and $r = 3$.
          $\binom{9}{2} \times (1 - 0.09)^7 \times 0.09^3 = 0.0136$

       (c) $\frac{1}{0.09} = 11.11$

       (d) $\frac{3}{0.09} = 33.33$

3.2.7   (a)   Consider a geometric distribution with parameter $p = 0.25$.
          $(1 - 0.25)^2 \times 0.25 = 0.1406$

       (b)   Consider a negative binomial distribution with parameters $p = 0.25$ and $r = 4$.
          $\binom{9}{3} \times (1 - 0.25)^6 \times 0.25^4 = 0.0584$

       The expected number of cards drawn before the fourth heart is obtained is the expectation of a negative binomial distribution with parameters $p = 0.25$ and $r = 4$, which is $\frac{4}{0.25} = 16$.

If the first two cards are spades then the probability that the first heart card is obtained on the fifth drawing is the same as the probability in part (a).

3.2.9   (a) Consider a geometric distribution with parameter $p = 0.6$.
$$P(X = 5) = (1 - 0.6)^4 \times 0.6 = 0.01536$$

  (b) Consider a negative binomial distribution with parameters $p = 0.6$ and $r = 4$.
$$P(X = 8) = \binom{7}{3} \times 0.6^4 \times 0.4^4 = 0.116$$

3.2.11 $P(X = 10) = \binom{9}{4} \times \left(\frac{1}{2}\right)^5 \times \left(\frac{1}{2}\right)^5 = 0.123$

3.2.13 $\frac{5}{0.65} = 7.69$

3.3 The Hypergeometric Distribution

3.3.1 (a) $P(X = 4) = \dfrac{\dbinom{6}{4} \times \dbinom{5}{3}}{\dbinom{11}{7}} = \dfrac{5}{11}$

(b) $P(X = 5) = \dfrac{\dbinom{6}{5} \times \dbinom{5}{2}}{\dbinom{11}{7}} = \dfrac{2}{11}$

(c) $P(X \leq 3) = P(X = 2) + P(X = 3) = \dfrac{23}{66}$

3.3.3 (a) $\dfrac{\dbinom{10}{3} \times \dbinom{7}{2}}{\dbinom{17}{5}} = \dfrac{90}{221}$

(b) $\dfrac{\dbinom{10}{1} \times \dbinom{7}{4}}{\dbinom{17}{5}} = \dfrac{25}{442}$

(c) $P(\text{no red balls}) + P(\text{one red ball}) + P(\text{two red balls}) = \dfrac{139}{442}$

3.3.5 $\dfrac{\dbinom{12}{3} \times \dbinom{40}{2}}{\dbinom{52}{5}} = \dfrac{55}{833}$

The number of picture cards X in a hand of 13 cards has a hypergeometric distribution with $N = 52$, $n = 13$, and $r = 12$.

The expected value is

$E(X) = \frac{13 \times 12}{52} = 3$

and the variance is

$\text{Var}(X) = \left(\frac{52-13}{52-1}\right) \times 13 \times \frac{12}{52} \times \left(1 - \frac{12}{52}\right) = \frac{30}{17}.$

3.3.7   (a) $\dfrac{\binom{7}{3} \times \binom{4}{0}}{\binom{11}{3}} = \frac{7}{33}$

(b) $\dfrac{\binom{7}{1} \times \binom{4}{2}}{\binom{11}{3}} = \frac{14}{55}$

3.3.9   (a) $\dfrac{\binom{8}{2} \times \binom{8}{2}}{\binom{16}{4}} = \frac{28}{65} = 0.431$

(b) $P\left(B\left(4, \frac{1}{2}\right) = 2\right) = \binom{4}{2} \times \left(\frac{1}{2}\right)^2 \times \left(\frac{1}{2}\right)^2 = \frac{3}{8} = 0.375$

3.3.11 $\quad P(2) = \dfrac{\dbinom{7}{2} \times \dbinom{3}{2}}{\dbinom{10}{4}} = \dfrac{3}{10}$

$$P(3) = \dfrac{\dbinom{7}{1} \times \dbinom{3}{3}}{\dbinom{10}{4}} = \dfrac{1}{30}$$

$$P(2) + P(3) = \dfrac{3}{10} + \dfrac{1}{30} = \dfrac{1}{3}$$

3.4   The Poisson Distribution

3.4.1   (a) $P(X = 1) = \frac{e^{-3.2} \times 3.2^1}{1!} = 0.1304$

(b) $P(X \leq 3) = P(X = 0) + P(X = 1) + P(X = 2) + P(X = 3) = 0.6025$

(c) $P(X \geq 6) = 1 - P(X \leq 5) = 0.1054$

(d) $P(X = 0 | X \leq 3) = \frac{P(X=0)}{P(X \leq 3)} = \frac{0.0408}{0.6025} = 0.0677$

3.4.5   It is best to use a Poisson distribution with $\lambda = \frac{25}{100} = 0.25$.

$P(X = 0) = \frac{e^{-0.25} \times 0.25^0}{0!} = 0.7788$

$P(X \leq 1) = P(X = 0) + P(X = 1) = 0.9735$

3.4.7   A $B(500, 0.005)$ distribution can be approximated by a

Poisson distribution with $\lambda = 500 \times 0.005 = 2.5$.

Therefore,

$P(B(500, 0.005) \leq 3)$

$\simeq \frac{e^{-2.5} \times 2.5^0}{0!} + \frac{e^{-2.5} \times 2.5^1}{1!} + \frac{e^{-2.5} \times 2.5^2}{2!} + \frac{e^{-2.5} \times 2.5^3}{3!}$

$= 0.7576$

3.4.9   B

3.5 The Multinomial Distribution

3.5.1   (a) $\frac{11!}{4! \times 5! \times 2!} \times 0.23^4 \times 0.48^5 \times 0.29^2 = 0.0416$

(b) $P(B(7, 0.23) < 3) = 0.7967$

3.5.3   (a) $\frac{8!}{2! \times 5! \times 1!} \times 0.09^2 \times 0.79^5 \times 0.12^1 = 0.0502$

(b) $\frac{8!}{1! \times 5! \times 2!} \times 0.09^1 \times 0.79^5 \times 0.12^2 = 0.0670$

(c) $P(B(8, 0.09) \geq 2) = 0.1577$

The expected number of misses is $8 \times 0.12 = 0.96$.

3.5.5   The probability that an order is received over the internet and it is large is $0.6 \times 0.3 = 0.18$.

The probability that an order is received over the internet and it is small is $0.6 \times 0.7 = 0.42$.

The probability that an order is not received over the internet and it is large is $0.4 \times 0.4 = 0.16$.

The probability that an order is not received over the internet and it is small is $0.4 \times 0.6 = 0.24$.

The answer is $\frac{8!}{2! \times 2! \times 2! \times 2!} \times 0.18^2 \times 0.42^2 \times 0.16^2 \times 0.24^2 = 0.0212$.

Chapter 4

Continuous Probability Distributions

4.1 The Uniform Distribution

4.1.1   (a) $E(X) = \frac{-3+8}{2} = 2.5$

         (b) $\sigma = \frac{8-(-3)}{\sqrt{12}} = 3.175$

         (c)   The upper quartile is 5.25.

         (d) $P(0 \leq X \leq 4) = \int_0^4 \frac{1}{11} \, dx = \frac{4}{11}$

4.1.3   (a)   These four intervals have probabilities 0.30, 0.20, 0.25, and 0.25 respectively, and the expectations and variances are calculated from the binomial distribution.

           The expectations are:

           $20 \times 0.30 = 6$

           $20 \times 0.20 = 4$

           $20 \times 0.25 = 5$

           $20 \times 0.25 = 5$

The variances are:

$20 \times 0.30 \times 0.70 = 4.2$

$20 \times 0.20 \times 0.80 = 3.2$

$20 \times 0.25 \times 0.75 = 3.75$

$20 \times 0.25 \times 0.75 = 3.75$

(b) Using the multinomial distribution the probability is

$\frac{20!}{5! \times 5! \times 5! \times 5!} \times 0.30^5 \times 0.20^5 \times 0.25^5 \times 0.25^5 = 0.0087.$

4.1.5   (a) The probability is $\frac{4.184 - 4.182}{4.185 - 4.182} = \frac{2}{3}$.

(b) $P(\text{difference} \leq 0.0005 \mid \text{fits in hole}) = \frac{P(4.1835 \leq \text{diameter} \leq 4.1840)}{P(\text{diameter} \leq 4.1840)}$

$= \frac{4.1840 - 4.1835}{4.1840 - 4.1820} = \frac{1}{4}$

4.2   The Exponential Distribution

4.2.3   (a) $E(X) = \frac{1}{0.2} = 5$

(b) $\sigma = \frac{1}{0.2} = 5$

(c) The median is $\frac{0.693}{0.2} = 3.47$.

(d) $P(X \geq 7) = 1 - F(7) = 1 - (1 - e^{-0.2 \times 7}) = e^{-1.4} = 0.2466$

(e) The memoryless property of the exponential distribution implies that the required probability is

$$P(X \geq 2) = 1 - F(2) = 1 - (1 - e^{-0.2 \times 2}) = e^{-0.4} = 0.6703.$$

4.2.5 $F(x) = \int_{-\infty}^{x} \frac{1}{2}\lambda e^{-\lambda(\theta - y)} \, dy = \frac{1}{2}e^{-\lambda(\theta - x)}$

for $-\infty \leq x \leq \theta$, and

$F(x) = \frac{1}{2} + \int_{\theta}^{x} \frac{1}{2}\lambda e^{-\lambda(y - \theta)} \, dy = 1 - \frac{1}{2}e^{-\lambda(x - \theta)}$

for $\theta \leq x \leq \infty$.

(a) $P(X \leq 0) = F(0) = \frac{1}{2}e^{-3(2-0)} = 0.0012$

(b) $P(X \geq 1) = 1 - F(1) = 1 - \frac{1}{2}e^{-3(2-1)} = 0.9751$

4.2.7   (a) $\lambda = 1.8$

(b) $E(X) = \frac{1}{1.8} = 0.5556$

(c) $P(X \geq 1) = 1 - F(1) = 1 - (1 - e^{-1.8 \times 1}) = e^{-1.8} = 0.1653$

(d) A Poisson distribution with parameter $1.8 \times 4 = 7.2$.

(e) $P(X \geq 4) = 1 - P(X = 0) - P(X = 1) - P(X = 2) - P(X = 3)$

$= 1 - \frac{e^{-7.2} \times 7.2^0}{0!} - \frac{e^{-7.2} \times 7.2^1}{1!} - \frac{e^{-7.2} \times 7.2^2}{2!} - \frac{e^{-7.2} \times 7.2^3}{3!} = 0.9281$

4.2.9   (a) $P(X \geq 1.5) = e^{-0.8 \times 1.5} = 0.301$

(b) The number of arrivals Y has a Poisson distribution with parameter
$0.8 \times 2 = 1.6$
so that the required probability is

$P(Y \geq 3) = 1 - P(Y = 0) - P(Y = 1) - P(Y = 2)$

$= 1 - \left(e^{-1.6} \times \frac{1.6^0}{0!}\right) - \left(e^{-1.6} \times \frac{1.6^1}{1!}\right) - \left(e^{-1.6} \times \frac{1.6^2}{2!}\right) = 0.217$

4.2.11   (a) $P(X \leq 6) = 1 - e^{-0.2 \times 6} = 0.699$

(b) The number of arrivals Y has a Poisson distribution with parameter
$0.2 \times 10 = 2$
so that the required probability is

$P(Y = 3) = e^{-2} \times \frac{2^3}{3!} = 0.180$

4.2.13   The number of signals X in a 100 meter stretch has a Poisson distribution
with mean $0.022 \times 100 = 2.2$.

$P(X \leq 1) = P(X = 0) + P(X = 1)$

$= \left(e^{-2.2} \times \frac{2.2^0}{0!}\right) + \left(e^{-2.2} \times \frac{2.2^1}{1!}\right)$

$= 0.111 + 0.244 = 0.355$

4.3   The Gamma Distribution

4.3.1 $\Gamma(5.5) = 4.5 \times 3.5 \times 2.5 \times 1.5 \times 0.5 \times \sqrt{\pi} = 52.34$

4.3.3   (a) $f(3) = 0.2055$
$F(3) = 0.3823$
$F^{-1}(0.5) = 3.5919$

       (b) $f(3) = 0.0227$
$F(3) = 0.9931$
$F^{-1}(0.5) = 1.3527$

       (c) $f(3) = 0.2592$
$F(3) = 0.6046$
$F^{-1}(0.5) = 2.6229$

In this case

$$f(3) = \frac{1.4^4 \times 3^{4-1} \times e^{-1.4 \times 3}}{3!} = 0.2592.$$

4.3.5   (a) A gamma distribution with parameters $k = 4$ and $\lambda = 2$.

       (b) $E(X) = \frac{4}{2} = 2$

       (c) $\sigma = \frac{\sqrt{4}}{2} = 1$

       (d) The probability can be calculated as
$P(X \geq 3) = 0.1512$
where the random variable X has a gamma distribution with parameters $k = 4$ and $\lambda = 2$.

The probability can also be calculated as
$P(Y \leq 3) = 0.1512$
where the random variable Y has a Poisson distribution with parameter $2 \times 3 = 6$ which counts the number of imperfections in a 3 meter length of fiber.

4.3.7   (a) The expectation is $E(X) = \frac{44}{0.7} = 62.86$

the variance is $\mathrm{Var}(X) = \frac{44}{0.7^2} = 89.80$

and the standard deviation is $\sqrt{89.80} = 9.48$.

(b) $F(60) = 0.3991$

4.4   The Weibull Distribution

4.4.3   (a) $\frac{(-\ln(1-0.5))^{1/2.3}}{1.7} = 0.5016$

     (b) $\frac{(-\ln(1-0.75))^{1/2.3}}{1.7} = 0.6780$

        $\frac{(-\ln(1-0.25))^{1/2.3}}{1.7} = 0.3422$

     (c) $F(x) = 1 - e^{-(1.7x)^{2.3}}$

        $P(0.5 \leq X \leq 1.5) = F(1.5) - F(0.5) = 0.5023$

4.4.5   (a) $\frac{(-\ln(1-0.5))^{1/0.4}}{0.5} = 0.8000$

     (b) $\frac{(-\ln(1-0.75))^{1/0.4}}{0.5} = 4.5255$

        $\frac{(-\ln(1-0.25))^{1/0.4}}{0.5} = 0.0888$

     (c) $\frac{(-\ln(1-0.95))^{1/0.4}}{0.5} = 31.066$

        $\frac{(-\ln(1-0.99))^{1/0.4}}{0.5} = 91.022$

     (d) $F(x) = 1 - e^{-(0.5x)^{0.4}}$

        $P(3 \leq X \leq 5) = F(5) - F(3) = 0.0722$

4.4.7   The probability that a culture has developed within four days is
$F(4) = 1 - e^{-(0.3 \times 4)^{0.6}} = 0.672.$

Using the negative binomial distribution, the probability that exactly ten cultures are opened is

$\binom{9}{4} \times (1 - 0.672)^5 \times 0.672^5 = 0.0656.$

4.5 The Beta Distribution

4.5.1 (a) Since

$$\int_0^1 A\, x^3(1-x)^2\, dx = 1$$

it follows that $A = 60$.

(b) $E(X) = \int_0^1 60\, x^4(1-x)^2\, dx = \frac{4}{7}$

$E(X^2) = \int_0^1 60\, x^5(1-x)^2\, dx = \frac{5}{14}$

Therefore,

$\mathrm{Var}(X) = \frac{5}{14} - \left(\frac{4}{7}\right)^2 = \frac{3}{98}$.

(c) This is a beta distribution with $a = 4$ and $b = 3$.

$E(X) = \frac{4}{4+3} = \frac{4}{7}$

$\mathrm{Var}(X) = \frac{4\times 3}{(4+3)^2 \times (4+3+1)} = \frac{3}{98}$

4.5.3 (a) $f(0.5) = 1.9418$
$F(0.5) = 0.6753$
$F^{-1}(0.75) = 0.5406$

(b) $f(0.5) = 0.7398$
$F(0.5) = 0.7823$
$F^{-1}(0.75) = 0.4579$

(c) $f(0.5) = 0.6563$
$F(0.5) = 0.9375$
$F^{-1}(0.75) = 0.3407$

In this case

$$f(0.5) = \frac{\Gamma(2+6)}{\Gamma(2)\times\Gamma(6)} \times 0.5^{2-1} \times (1-0.5)^{6-1} = 0.65625.$$

4.5.5   (a) $E(X) = \frac{7.2}{7.2+2.3} = 0.7579$

$\text{Var}(X) = \frac{7.2 \times 2.3}{(7.2+2.3)^2 \times (7.2+2.3+1)} = 0.0175$

(b) From the computer $P(X \geq 0.9) = 0.1368$.

Chapter 5

The Normal Distribution

5.1 Probability Calculations

5.1.1   (a) $\Phi(1.34) = 0.9099$

      (b) $1 - \Phi(-0.22) = 0.5871$

      (c) $\Phi(0.43) - \Phi(-2.19) = 0.6521$

      (d) $\Phi(1.76) - \Phi(0.09) = 0.4249$

      (e) $\Phi(0.38) - \Phi(-0.38) = 0.2960$

      (f) Solving $\Phi(x) = 0.55$ gives $x = 0.1257$.

      (g) Solving $1 - \Phi(x) = 0.72$ gives $x = -0.5828$.

      (h) Solving $\Phi(x) - \Phi(-x) = (2 \times \Phi(x)) - 1 = 0.31$ gives $x = 0.3989$.

5.1.3   (a) $P(X \leq 10.34) = \Phi\left(\frac{10.34-10}{\sqrt{2}}\right) = 0.5950$

      (b) $P(X \geq 11.98) = 1 - \Phi\left(\frac{11.98-10}{\sqrt{2}}\right) = 0.0807$

69

(c) $P(7.67 \leq X \leq 9.90) = \Phi\left(\frac{9.90-10}{\sqrt{2}}\right) - \Phi\left(\frac{7.67-10}{\sqrt{2}}\right) = 0.4221$

(d) $P(10.88 \leq X \leq 13.22) = \Phi\left(\frac{13.22-10}{\sqrt{2}}\right) - \Phi\left(\frac{10.88-10}{\sqrt{2}}\right) = 0.2555$

(e) $P(|X - 10| \leq 3) = P(7 \leq X \leq 13)$

$\quad = \Phi\left(\frac{13-10}{\sqrt{2}}\right) - \Phi\left(\frac{7-10}{\sqrt{2}}\right) = 0.9662$

(f) Solving $P(N(10,2) \leq x) = 0.81$ gives $x = 11.2415$.

(g) Solving $P(N(10,2) \geq x) = 0.04$ gives $x = 12.4758$.

(h) Solving $P(|N(10,2) - 10| \geq x) = 0.63$ gives $x = 0.6812$.

5.1.5   Solving

$P(X \leq 5) = \Phi\left(\frac{5-\mu}{\sigma}\right) = 0.8$

and

$P(X \geq 0) = 1 - \Phi\left(\frac{0-\mu}{\sigma}\right) = 0.6$

gives $\mu = 1.1569$ and $\sigma = 4.5663$.

5.1.7 $P(X \leq \mu + \sigma z_\alpha) = \Phi\left(\frac{\mu + \sigma z_\alpha - \mu}{\sigma}\right)$

$= \Phi(z_\alpha) = 1 - \alpha$

$P(\mu - \sigma z_{\alpha/2} \leq X \leq \mu + \sigma z_{\alpha/2}) = \Phi\left(\frac{\mu + \sigma z_{\alpha/2} - \mu}{\sigma}\right) - \Phi\left(\frac{\mu - \sigma z_{\alpha/2} - \mu}{\sigma}\right)$

$= \Phi(z_{\alpha/2}) - \Phi(-z_{\alpha/2})$

$= 1 - \alpha/2 - \alpha/2 = 1 - \alpha$

5.1.9   (a) $P(N(3.00, 0.12^2) \geq 3.2) = 0.0478$

(b) $P(N(3.00, 0.12^2) \leq 2.7) = 0.0062$

(c) Solving
$$P(3.00 - c \leq N(3.00, 0.12^2) \leq 3.00 + c) = 0.99$$
gives
$$c = 0.12 \times z_{0.005} = 0.12 \times 2.5758 = 0.3091.$$

5.1.11 (a) Solving $P(N(4.3, 0.12^2) \leq x) = 0.75$ gives $x = 4.3809$.
Solving $P(N(4.3, 0.12^2) \leq x) = 0.25$ gives $x = 4.2191$.

(b) Solving
$$P(4.3 - c \leq N(4.3, 0.12^2) \leq 4.3 + c) = 0.80$$
gives
$$c = 0.12 \times z_{0.10} = 0.12 \times 1.2816 = 0.1538.$$

5.1.13 (a) $P(N(23.8, 1.28) \leq 23.0) = 0.2398$

(b) $P(N(23.8, 1.28) \geq 24.0) = 0.4298$

(c) $P(24.2 \leq N(23.8, 1.28) \leq 24.5) = 0.0937$

(d) Solving $P(N(23.8, 1.28) \leq x) = 0.75$ gives $x = 24.56$.

(e) Solving $P(N(23.8, 1.28) \leq x) = 0.95$ gives $x = 25.66$.

5.1.15 (a) $P(2599 \leq X \leq 2601) = \Phi\left(\frac{2601 - 2600}{0.6}\right) - \Phi\left(\frac{2599 - 2600}{0.6}\right)$
$$= 0.9522 - 0.0478 = 0.9044$$

The probability of being outside the range is $1 - 0.9044 = 0.0956$.

(b) It is required that
$$P(2599 \leq X \leq 2601) = \Phi\left(\frac{2601 - 2600}{\sigma}\right) - \left(\frac{2599 - 2600}{\sigma}\right)$$
$$= 1 - 0.001 = 0.999.$$

Consequently,

$$\Phi\left(\frac{1}{\sigma}\right) - \Phi\left(\frac{-1}{\sigma}\right)$$

$$= 2\Phi\left(\frac{1}{\sigma}\right) - 1 = 0.999$$

so that

$$\Phi\left(\frac{1}{\sigma}\right) = 0.9995$$

which gives

$$\frac{1}{\sigma} = \Phi^{-1}(0.9995) = 3.2905$$

with

$$\sigma = 0.304.$$

5.1.17 $0.95 = P(N(\mu, 4.2^2) \le 100) = P\left(N(0,1) \le \frac{100-\mu}{4.2}\right)$

Therefore,

$$\frac{100-\mu}{4.2} = z_{0.05} = 1.645$$

so that $\mu = 93.09$.

5.2   Linear Combinations of Normal Random Variables

5.2.1   (a) $P(N(3.2 + (-2.1), 6.5 + 3.5) \geq 0) = 0.6360$

        (b) $P(N(3.2 + (-2.1) - (2 \times 12.0), 6.5 + 3.5 + (2^2 \times 7.5)) \leq -20) = 0.6767$

        (c) $P(N((3 \times 3.2) + (5 \times (-2.1)), (3^2 \times 6.5) + (5^2 \times 3.5)) \geq 1) = 0.4375$

        (d)   The mean is $(4 \times 3.2) - (4 \times (-2.1)) + (2 \times 12.0) = 45.2$.
             The variance is $(4^2 \times 6.5) + (4^2 \times 3.5) + (2^2 \times 7.5) = 190$.
             $P(N(45.2, 190) \leq 25) = 0.0714$

        (e) $P(|\ N(3.2 + (6 \times (-2.1)) + 12.0, 6.5 + (6^2 \times 3.5) + 7.5)\ | \geq 2) = 0.8689$

        (f) $P(|\ N((2 \times 3.2) - (-2.1) - 6, (2^2 \times 6.5) + 3.5)\ | \leq 1) = 0.1315$

5.2.3   (a) $\Phi(0.5) - \Phi(-0.5) = 0.3830$

        (b) $P\left(\left|N\left(0, \frac{1}{8}\right)\right| \leq 0.5\right) = 0.8428$

        (c)   It is required that

             $0.5\sqrt{n} \geq z_{0.005} = 2.5758$

            which is satisfied for $n \geq 27$.

5.2.5 $P(144 \leq N(37 + 37 + 24 + 24 + 24, 0.49 + 0.49 + 0.09 + 0.09 + 0.09)$
        $\leq 147) = 0.7777$

5.2.7   (a) $1.05y + 1.05(1000 - y) = \$1050$

        (b) $0.0002y^2 + 0.0003(1000 - y)^2$

        (c)   The variance is minimized with $y = 600$ and the minimum variance
            is 120.

$$P(N(1050, 120) \geq 1060) = 0.1807$$

5.2.9   (a) $N(22 \times 1.03, 22 \times 0.014^2) = N(22.66, 4.312 \times 10^{-3})$

(b) Solving $P(N(22.66, 4.312 \times 10^{-3}) \leq x) = 0.75$ gives $x = 22.704$.
     Solving $P(N(22.66, 4.312 \times 10^{-3}) \leq x) = 0.25$ gives $x = 22.616$.

5.2.11   (a) $P\left(4.2 \leq N\left(4.5, \frac{0.88}{15}\right) \leq 4.9\right)$

$$= P\left(\frac{\sqrt{15}(4.2-4.5)}{\sqrt{0.88}} \leq N(0,1) \leq \frac{\sqrt{15}(4.9-4.5)}{\sqrt{0.88}}\right)$$

$$= \Phi(1.651) - \Phi(-1.239)$$

$$= 0.951 - 0.108 = 0.843$$

(b) $0.99 = P\left(4.5 - c \leq N\left(4.5, \frac{0.88}{15}\right) \leq 4.5 + c\right)$

$$= P\left(\frac{-c\sqrt{15}}{\sqrt{0.88}} \leq N(0,1) \leq \frac{c\sqrt{15}}{\sqrt{0.88}}\right)$$

Therefore,

$$\frac{c\sqrt{15}}{\sqrt{0.88}} = z_{0.005} = 2.576$$

so that $c = 0.624$.

5.2.13   The height of a stack of 4 components of type A has a normal distribution
with mean $4 \times 190 = 760$ and a standard deviation $\sqrt{4} \times 10 = 20$.

The height of a stack of 5 components of type B has a normal distribution
with mean $5 \times 150 = 750$ and a standard deviation $\sqrt{5} \times 8 = 17.89$.

$$P(N(760, 20^2) > N(750, 17.89^2))$$

$$= P(N(760 - 750, 20^2 + 17.78^2) > 0)$$

$$= P\left(N(0,1) > \frac{-10}{\sqrt{720}}\right)$$

$$= 1 - \Phi(-0.373) = 0.645$$

5.2.15   It is required that

$$P\left(N\left(110, \tfrac{4}{n}\right) \le 111\right)$$

$$= P\left(N(0,1) \le \tfrac{\sqrt{n}(111-110)}{2}\right) \ge 0.99.$$

Therefore,

$$\tfrac{\sqrt{n}(111-110)}{2} \ge z_{0.01} = 2.326$$

which is satisfied for $n \ge 22$.

5.2.17   (a) $E(X) = 20\mu = 20 \times 63400 = 1268000$
The standard deviation is $\sqrt{20}\,\sigma = \sqrt{20} \times 2500 = 11180.$

(b) $E(X) = \mu = 63400$
The standard deviation is $\tfrac{\sigma}{\sqrt{30}} = \tfrac{2500}{\sqrt{30}} = 456.4.$

5.2.19 $P\left(N\left(30000 + 45000, 4000^2 + 3000^2\right) \ge 85000\right)$
$= P\left(N\left(0,1\right) \ge 2\right) = 0.023$

5.3   Approximating Distributions

5.3.1   (a)   The exact probability is 0.3823.

The normal approximation is

$$1 - \Phi\left(\frac{8-0.5-(10\times0.7)}{\sqrt{10\times0.7\times0.3}}\right) = 0.3650.$$

(b)   The exact probability is 0.9147.

The normal approximation is

$$\Phi\left(\frac{7+0.5-(15\times0.3)}{\sqrt{15\times0.3\times0.7}}\right) - \Phi\left(\frac{1+0.5-(15\times0.3)}{\sqrt{15\times0.3\times0.7}}\right) = 0.9090.$$

(c)   The exact probability is 0.7334.

The normal approximation is

$$\Phi\left(\frac{4+0.5-(9\times0.4)}{\sqrt{9\times0.4\times0.6}}\right) = 0.7299.$$

(d)   The exact probability is 0.6527.

The normal approximation is

$$\Phi\left(\frac{11+0.5-(14\times0.6)}{\sqrt{14\times0.6\times0.4}}\right) - \Phi\left(\frac{7+0.5-(14\times0.6)}{\sqrt{14\times0.6\times0.4}}\right) = 0.6429.$$

5.3.3   The required probability is

$$\Phi\left(0.02\sqrt{n} + \tfrac{1}{\sqrt{n}}\right) - \Phi\left(-0.02\sqrt{n} - \tfrac{1}{\sqrt{n}}\right)$$

which is equal to

0.2358 for $n = 100$

0.2764 for $n = 200$

0.3772 for $n = 500$

0.4934 for $n = 1000$

and 0.6408 for $n = 2000$.

5.3.5   (a) A normal distribution can be used with

$$\mu = 500 \times 2.4 = 1200$$

and

$$\sigma^2 = 500 \times 2.4 = 1200.$$

  (b) $P(N(1200, 1200) \geq 1250) = 0.0745$

5.3.7   The normal approximation is

$$\Phi\left(\frac{200 + 0.5 - (250{,}000 \times 0.0007)}{\sqrt{250{,}000 \times 0.0007 \times 0.9993}}\right) = 0.9731.$$

5.3.9   The yearly income can be approximated by a normal distribution with

$$\mu = 365 \times \frac{5}{0.9} = 2027.8$$

and

$$\sigma^2 = 365 \times \frac{5}{0.9^2} = 2253.1.$$

$$P(N(2027.8, 2253.1) \geq 2000) = 0.7210$$

5.3.11   The expectation of the strength of a chemical solution is

$$E(X) = \frac{18}{18 + 11} = 0.6207$$

and the variance is

$$\mathrm{Var}(X) = \frac{18 \times 11}{(18+11)^2(18+11+1)} = 0.007848.$$

Using the central limit theorem the required probability can be estimated as

$$P\left(0.60 \leq N\left(0.6207, \frac{0.007848}{20}\right) \leq 0.65\right)$$

$$= \Phi(1.479) - \Phi(-1.045) = 0.7824.$$

5.3.13 $P(60 \leq X \leq 100) = (1 - e^{-100/84}) - (1 - e^{-60/84}) = 0.1855$

$P(B(350, 0.1855) \geq 55)$

$\simeq P(N(350 \times 0.1855, 350 \times 0.1855 \times 0.8145) \geq 54.5)$

$= P\left(N(0,1) \geq \frac{54.5 - 64.925}{7.272}\right)$

$= 1 - \Phi(-1.434) = 0.9239$

5.3.15   (a) $P(X \geq 891.2) = \frac{892 - 891.2}{892 - 890} = 0.4$

Using the negative binomial distribution the required probability is

$\binom{5}{2} \times 0.4^3 \times 0.6^3 = 0.1382.$

(b) $P(X \geq 890.7) = \frac{892 - 890.7}{892 - 890} = 0.65$

$P(B(200, 0.65) \geq 100)$

$\simeq P(N(200 \times 0.65, 200 \times 0.65 \times 0.35) \geq 99.5)$

$= P\left(N(0,1) \geq \frac{99.5 - 130}{\sqrt{45.5}}\right)$

$= 1 - \Phi(-4.52) \simeq 1$

5.4 Distributions Related to the Normal Distribution

5.4.1   (a) $E(X) = e^{1.2+(1.5^2/2)} = 10.23$

        (b) $\text{Var}(X) = e^{(2\times 1.2)+1.5^2} \times (e^{1.5^2} - 1) = 887.69$

        (c)   Since $z_{0.25} = 0.6745$ the upper quartile is

            $e^{1.2+(1.5\times 0.6745)} = 9.13.$

        (d)   The lower quartile is

            $e^{1.2+(1.5\times(-0.6745))} = 1.21.$

        (e)   The interquartile range is $9.13 - 1.21 = 7.92.$

        (f) $P(5 \leq X \leq 8) = \Phi\left(\frac{\ln(8)-1.2}{1.5}\right) - \Phi\left(\frac{\ln(5)-1.2}{1.5}\right) = 0.1136$

5.4.5   (a) $\chi^2_{0.10,9} = 14.68$

        (b) $\chi^2_{0.05,20} = 31.41$

        (c) $\chi^2_{0.01,26} = 45.64$

        (d) $\chi^2_{0.90,50} = 37.69$

        (e) $\chi^2_{0.95,6} = 1.635$

5.4.7   (a) $t_{0.10,7} = 1.415$

        (b) $t_{0.05,19} = 1.729$

        (c) $t_{0.01,12} = 2.681$

        (d) $t_{0.025,30} = 2.042$

(e) $t_{0.005,4} = 4.604$

5.4.9   (a) $F_{0.10,9,10} = 2.347$

(b) $F_{0.05,6,20} = 2.599$

(c) $F_{0.01,15,30} = 2.700$

(d) $F_{0.05,4,8} = 3.838$

(e) $F_{0.01,20,13} = 3.665$

5.4.11   This follows from the definitions

$$t_\nu \sim \frac{N(0,1)}{\sqrt{\chi_\nu^2/\nu}}$$

and

$$F_{1,\nu} \sim \frac{\chi_1^2}{\chi_\nu^2/\nu}.$$

5.4.13 $P(F_{5,20} \geq 4.00) = 0.011$

5.4.15   (a) $P(F_{10,50} \geq 2.5) = 0.016$

(b) $P(\chi_{17}^2 \leq 12) = 0.200$

(c) $P(t_{24} \geq 3) = 0.003$

(d) $P(t_{14} \geq -2) = 0.967$

5.4.17   (a) $P(t_{16} \leq 1.9) = 0.962$

(b) $P(\chi_{25}^2 \geq 42.1) = 0.018$

(c) $P(F_{9,14} \leq 1.8) = 0.844$

(d) $P(-1.4 \leq t_{29} \leq 3.4) = 0.913$

Chapter 6

Descriptive Statistics

6.1   Experimentation

6.1.1   For this problem the population is the somewhat imaginary concept of "all possible die rolls."

The sample should be representative if the die is shaken properly.

6.1.3   Is the population all students? - or the general public? - or perhaps it should just be computing students at that college?

You have to consider whether the eye colors of computing students are representative of the eye colors of all students or of all people.

Perhaps eye colors are affected by race and the racial make-up of the class may not reflect that of the student body or the general public as a whole.

6.1.5   The population is all peach boxes received by the supermarket within the time period.

The random sampling within each day's shipment and the recording of an observation every day should ensure that the sample is reasonably representative.

6.1.7   The population may be all bricks shipped by that company, or just the bricks in that particular delivery.

82

The random selection of the sample should ensure that it is representative of that particular delivery of bricks.

However, that specific delivery of bricks may not be representative of all of the deliveries from that company.

6.1.9   The population is all plastic panels made by the machine.

If the 80 sample panels are selected in some random manner then they should be representative of the entire population.

6.2   Data Presentation

6.2.3   The smallest observation 1.097 and the largest observation 1.303 both appear to be outliers.

6.2.5   There would appear to be no reason to doubt that the die is a fair one.

A test of the fairness of the die could be made using the methods presented in section 10.3.

6.2.7   The assignment "other" is employed considerably less frequently than blue, green, and brown, which are each about equally frequent.

6.2.9   The observations 25 and 14 can be considered to be outliers.

6.2.11   The smallest observation 0.874 can be considered to be an outlier.

6.2.13   This is a negatively skewed data set.

The smallest observations 6.00 and 6.04 can be considered to be outliers, and possibly some of the other small observations may also be considered to be outliers.

6.3 Sample Statistics

Note: The sample statistics for the problems in this section depend upon whether any observations have been removed as outliers. To avoid confusion, the answers given here assume that **no** observations have been removed.

The trimmed means given here are those obtained by removing the largest 5% and the smallest 5% of the data observations.

6.3.1   The sample mean is $\bar{x} = 155.95$.

The sample median is 159.

The sample trimmed mean is 156.50.

The sample standard deviation is $s = 18.43$.

The upper sample quartile is 169.5.

The lower sample quartile is 143.25.

6.3.3   The sample mean is $\bar{x} = 37.08$.

The sample median is 35.

The sample trimmed mean is 36.35.

The sample standard deviation is $s = 8.32$.

The upper sample quartile is 40.

The lower sample quartile is 33.5.

6.3.5   The sample mean is $\bar{x} = 69.35$.

The sample median is 66.

The sample trimmed mean is 67.88.

The sample standard deviation is $s = 17.59$.

The upper sample quartile is 76.

The lower sample quartile is 61.

6.3.7   The sample mean is $\bar{x} = 12.211$.

          The sample median is 12.

          The sample trimmed mean is 12.175.

          The sample standard deviation is $s = 2.629$.

          The upper sample quartile is 14.

          The lower sample quartile is 10.

6.3.9   The sample mean is $\bar{x} = 0.23181$.

          The sample median is 0.220.

          The sample trimmed mean is 0.22875.

          The sample standard deviation is $s = 0.07016$.

          The upper sample quartile is 0.280.

          The lower sample quartile is 0.185.

6.3.11   The sample mean is

          $$\frac{65+x}{6}$$

          and

          $$\sum_{i=1}^{6} x_i^2 = 1037 + x^2.$$

          Therefore,

          $$s^2 = \frac{1037 + x^2 - (65+x)^2/6}{5}$$

          which by differentiation can be shown to be minimized when $x = 13$
          (which is the average of the other five data points).

6.3.13   B

6.3.15   A

Chapter 7

Statistical Estimation

7.2 Properties of Point Estimates

7.2.1   (a) $\text{bias}(\hat{\mu}_1) = 0$

         The point estimate $\hat{\mu}_1$ is unbiased.

         $\text{bias}(\hat{\mu}_2) = 0$
         The point estimate $\hat{\mu}_2$ is unbiased.

         $\text{bias}(\hat{\mu}_3) = 9 - \frac{\mu}{2}$

     (b) $\text{Var}(\hat{\mu}_1) = 6.2500$

         $\text{Var}(\hat{\mu}_2) = 9.0625$

         $\text{Var}(\hat{\mu}_3) = 1.9444$

         The point estimate $\hat{\mu}_3$ has the smallest variance.

     (c) $\text{MSE}(\hat{\mu}_1) = 6.2500$

         $\text{MSE}(\hat{\mu}_2) = 9.0625$

         $\text{MSE}(\hat{\mu}_3) = 1.9444 + (9 - \frac{\mu}{2})^2$
         This is equal to 26.9444 when $\mu = 8$.

7.2.3   (a) $\text{Var}(\hat{\mu}_1) = 2.5$

87

(b) The value $p = 0.6$ produces the smallest variance which is
$\text{Var}(\hat{\mu}) = 2.4$.

(c) The relative efficiency is $\frac{2.4}{2.5} = 0.96$.

7.2.5   (a) $a_1 + \ldots + a_n = 1$

(b) $a_1 = \ldots = a_n = \frac{1}{n}$

7.2.7 $\text{bias}(\hat{\mu}) = \frac{\mu_0 - \mu}{2}$

$\text{Var}(\hat{\mu}) = \frac{\sigma^2}{4}$

$\text{MSE}(\hat{\mu}) = \frac{\sigma^2}{4} + \frac{(\mu_0 - \mu)^2}{4}$

$\text{MSE}(X) = \sigma^2$

7.2.9 $\text{Var}\left(\frac{X_1 + X_2}{2}\right)$

$= \frac{\text{Var}(X_1) + \text{Var}(X_2)}{4}$

$= \frac{5.39^2 + 9.43^2}{4}$

$= 29.49$

The standard deviation is $\sqrt{29.49} = 5.43$.

7.3 Sampling Distributions

7.3.1 $\operatorname{Var}\left(\frac{X_1}{n_1}\right) = \frac{p(1-p)}{n_1}$

$\operatorname{Var}\left(\frac{X_2}{n_2}\right) = \frac{p(1-p)}{n_2}$

The relative efficiency is the ratio of these two variances which is $\frac{n_1}{n_2}$.

7.3.3 (a) $P\left(\left|N\left(0, \frac{7}{15}\right)\right| \leq 0.4\right) = 0.4418$

(b) $P\left(\left|N\left(0, \frac{7}{50}\right)\right| \leq 0.4\right) = 0.7150$

7.3.5 (a) Solving

$$P\left(32 \times \frac{\chi_{20}^2}{20} \leq c\right) = P\left(\chi_{20}^2 \leq \frac{5c}{8}\right) = 0.90$$

gives $c = 45.46$.

(b) Solving

$$P\left(32 \times \frac{\chi_{20}^2}{20} \leq c\right) = P\left(\chi_{20}^2 \leq \frac{5c}{8}\right) = 0.95$$

gives $c = 50.26$.

7.3.7 (a) Solving

$$P\left(\frac{|t_{20}|}{\sqrt{21}} \leq c\right) = 0.95$$

gives $c = \frac{t_{0.025,20}}{\sqrt{21}} = 0.4552$.

(b) Solving

$$P\left(\frac{|t_{20}|}{\sqrt{21}} \leq c\right) = 0.99$$

gives $c = \frac{t_{0.005,20}}{\sqrt{21}} = 0.6209$.

7.3.9 $\hat{\mu} = \bar{x} = 974.3$

$\text{s.e.}(\hat{\mu}) = \frac{s}{\sqrt{n}} = \sqrt{\frac{452.1}{35}} = 3.594$

7.3.11 $\hat{p} = \frac{33}{150} = 0.22$

$\text{s.e.}(\hat{p}) = \sqrt{\frac{\hat{p}\,(1-\hat{p})}{n}} = \sqrt{\frac{0.22 \times 0.78}{150}} = 0.0338$

7.3.13 $\hat{\mu} = \bar{x} = 69.35$

$\text{s.e.}(\hat{\mu}) = \frac{s}{\sqrt{n}} = \frac{17.59}{\sqrt{200}} = 1.244$

7.3.15 $\hat{\mu} = \bar{x} = 12.211$

$\text{s.e.}(\hat{\mu}) = \frac{s}{\sqrt{n}} = \frac{2.629}{\sqrt{90}} = 0.277$

7.3.17 $\hat{\mu} = \bar{x} = 0.23181$

$\text{s.e.}(\hat{\mu}) = \frac{s}{\sqrt{n}} = \frac{0.07016}{\sqrt{75}} = 0.00810$

7.3.19 If a sample of size $n = 100$ is used, then the probability is

$P(0.24 - 0.05 \leq \hat{p} \leq 0.24 + 0.05) = P(19 \leq B(100, 0.24) \leq 29).$

Using a normal approximation this can be estimated as

$\Phi\left(\frac{29 + 0.5 - 100 \times 0.24}{\sqrt{100 \times 0.24 \times 0.76}}\right) - \Phi\left(\frac{19 - 0.5 - 100 \times 0.24}{\sqrt{100 \times 0.24 \times 0.76}}\right)$

$= \Phi(1.288) - \Phi(-1.288) = 0.8022.$

If a sample of size $n = 200$ is used, then the probability is

$P(38 \leq B(200, 0.24) \leq 58).$

Using a normal approximation this can be estimated as

$\Phi\left(\frac{58 + 0.5 - 200 \times 0.24}{\sqrt{200 \times 0.24 \times 0.76}}\right) - \Phi\left(\frac{38 - 0.5 - 200 \times 0.24}{\sqrt{200 \times 0.24 \times 0.76}}\right)$

$$= \Phi(1.738) - \Phi(-1.738) = 0.9178.$$

7.3.21　$P(0.62 \leq \hat{p} \leq 0.64)$

$$= P(300 \times 0.62 \leq B(300, 0.63) \leq 300 \times 0.64)$$

$$\simeq P(185.5 \leq N(300 \times 0.63, 300 \times 0.63 \times 0.37) \leq 192.5)$$

$$= P\left(\frac{185.5 - 189}{\sqrt{69.93}} \leq N(0,1) \leq \frac{192.5 - 189}{\sqrt{69.93}}\right)$$

$$= \Phi(0.419) - \Phi(-0.419) = 0.324$$

7.3.23　$\sqrt{\frac{0.126 \times 0.874}{360}} = 0.017$

7.3.25　$P\left(\mu - 2 \leq N\left(\mu, \frac{5.2^2}{18}\right) \leq \mu + 2\right)$

$$= P\left(\frac{-\sqrt{18} \times 2}{5.2} \leq N(0.1) \leq \frac{\sqrt{18} \times 2}{5.2}\right)$$

$$= \Phi(1.632) - \Phi(-1.632) = 0.103$$

7.3.27　$P(X \geq 60) = e^{-0.02 \times 60} = 0.301$

Let Y be the number of components that last longer than one hour.

$$P\left(0.301 - 0.05 \leq \frac{Y}{110} \leq 0.301 + 0.05\right)$$

$$= P(27.6 \leq Y \leq 38.6)$$

$$= P(28 \leq B(110, 0.301) \leq 38)$$

$$\simeq P\left(27.5 \leq N(110 \times 0.301, 110 \times 0.301 \times 0.699) \leq 38.5\right)$$

$$= P\left(\frac{27.5 - 33.11}{\sqrt{23.14}} \leq N(0,1) \leq \frac{38.5 - 33.11}{\sqrt{23.14}}\right)$$

$$= \Phi(1.120) - \Phi(-1.166)$$

$$= 0.869 - 0.122 = 0.747$$

7.3.29   (a) $p = \frac{592}{3288} = 0.18$

$$P(p - 0.1 \le \hat{p} \le p + 0.1)$$

$$= P\left(0.08 \le \frac{X}{20} \le 0.28\right)$$

$$= P(1.6 \le X \le 5.6)$$

where $X \sim B(20, 0.18)$.

This probability is

$$P(X = 2) + P(X = 3) + P(X = 4) + P(X = 5)$$

$$= \binom{20}{2} \times 0.18^2 \times 0.82^{18} + \binom{20}{3} \times 0.18^3 \times 0.82^{17}$$

$$+ \binom{20}{4} \times 0.18^4 \times 0.82^{16} + \binom{20}{5} \times 0.18^5 \times 0.82^{15}$$

$$= 0.7626.$$

(b) The probability that a sampled meter is operating outside the acceptable tolerance limits is now

$p^* = \frac{184}{2012} = 0.09.$

$$P(p - 0.1 \le \hat{p} \le p + 0.1)$$

$$= P\left(0.08 \le \frac{Y}{20} \le 0.28\right)$$

$$= P(1.6 \le Y \le 5.6)$$

where $Y \sim B(20, 0.09)$.

This probability is

$$P(Y = 2) + P(Y = 3) + P(Y = 4) + P(Y = 5)$$

$$= \binom{20}{2} \times 0.09^2 \times 0.91^{18} + \binom{20}{3} \times 0.09^3 \times 0.91^{17}$$

$$+ \binom{20}{4} \times 0.09^4 \times 0.91^{16} + \binom{20}{5} \times 0.09^5 \times 0.91^{15}$$

$$= 0.5416.$$

7.3.31   D

7.3.33   B

7.3.35   A

7.4   Constructing Parameter Estimates

7.4.1 $\hat{\lambda} = \bar{x} = 5.63$

$$\text{s.e.}(\hat{\lambda}) = \sqrt{\frac{\hat{\lambda}}{n}} = \sqrt{\frac{5.63}{23}} = 0.495$$

7.4.3   Using the method of moments

$$E(X) = \frac{1}{\lambda} = \bar{x}$$

which gives $\hat{\lambda} = \frac{1}{\bar{x}}$.

The likelihood is

$$L(x_1, \ldots, x_n, \lambda) = \lambda^n \, e^{-\lambda(x_1 + \ldots + x_n)}$$

which is maximized at $\hat{\lambda} = \frac{1}{\bar{x}}$.

7.4.5   Using the method of moments

$$E(X) = \frac{5}{\lambda} = \bar{x}$$

which gives $\hat{\lambda} = \frac{5}{\bar{x}}$.

The likelihood is

$$L(x_1, \ldots, x_n, \lambda) = \left(\frac{1}{24}\right)^n \times \lambda^{5n} \times x_1^4 \times \ldots \times x_n^4 \times e^{-\lambda(x_1 + \ldots + x_n)}$$

which is maximized at $\hat{\lambda} = \frac{5}{\bar{x}}$.

Chapter 8

Inferences on a Population Mean

8.1 Confidence Intervals

8.1.1 With $t_{0.025,30} = 2.042$ the confidence interval is

$$\left(53.42 - \frac{2.042 \times 3.05}{\sqrt{31}}, 53.42 + \frac{2.042 \times 3.05}{\sqrt{31}}\right) = (52.30, 54.54).$$

8.1.3 At 90% confidence the critical point is $t_{0.05,19} = 1.729$ and the confidence interval is

$$\left(436.5 - \frac{1.729 \times 11.90}{\sqrt{20}}, 436.5 + \frac{1.729 \times 11.90}{\sqrt{20}}\right) = (431.9, 441.1).$$

At 95% confidence the critical point is $t_{0.025,19} = 2.093$ and the confidence interval is

$$\left(436.5 - \frac{2.093 \times 11.90}{\sqrt{20}}, 436.5 + \frac{2.093 \times 11.90}{\sqrt{20}}\right) = (430.9, 442.1).$$

At 99% confidence the critical point is $t_{0.005,19} = 2.861$ and the confidence interval is

$$\left(436.5 - \frac{2.861 \times 11.90}{\sqrt{20}}, 436.5 + \frac{2.861 \times 11.90}{\sqrt{20}}\right) = (428.9, 444.1).$$

Even the 99% confidence level confidence interval does not contain the

95

value 450.0, and so 450.0 is not a plausible value for the average breaking strength.

8.1.5   With $z_{0.025} = 1.960$ the confidence interval is

$$\left(0.0328 - \tfrac{1.960 \times 0.015}{\sqrt{28}}, 0.0328 + \tfrac{1.960 \times 0.015}{\sqrt{28}}\right) = (0.0272, 0.0384).$$

8.1.7   With $t_{0.025, n-1} \simeq 2.0$ a sufficient sample size can be estimated as

$$n \geq 4 \times \left(\frac{t_{0.025, n-1}\, s}{L_0}\right)^2$$

$$= 4 \times \left(\frac{2.0 \times 10.0}{5}\right)^2 = 64.$$

A sample size of about $n = 64$ should be sufficient.

8.1.9   A total sample size of

$$n \geq 4 \times \left(\frac{t_{0.025, n_1 - 1}\, s}{L_0}\right)^2$$

$$= 4 \times \left(\frac{2.042 \times 3.05}{2.0}\right)^2 = 38.8$$

is required.

Therefore, an additional sample of at least $39 - 31 = 8$ observations should be sufficient.

8.1.11   A total sample size of

$$n \geq 4 \times \left(\frac{t_{0.005, n_1 - 1}\, s}{L_0}\right)^2$$

$$= 4 \times \left(\frac{2.861 \times 11.90}{10.0}\right)^2 = 46.4$$

is required.

Therefore, an additional sample of at least $47 - 20 = 27$ observations should be sufficient.

8.1.13   With $t_{0.01,60} = 2.390$ the value of c is obtained as

$$c = \bar{x} - \frac{t_{\alpha,n-1}\,s}{\sqrt{n}} = 0.768 - \frac{2.390 \times 0.0231}{\sqrt{61}} = 0.761.$$

The confidence interval contains the value 0.765, and so it is plausible that the average solution density is less than 0.765.

8.1.15   With $z_{0.01} = 2.326$ the value of c is obtained as

$$c = \bar{x} + \frac{z_{\alpha}\,\sigma}{\sqrt{n}} = 415.7 + \frac{2.326 \times 10.0}{\sqrt{29}} = 420.0.$$

The confidence interval contains the value 418.0, and so it is plausible that the mean radiation level is greater than 418.0.

8.1.17   Using the critical point $t_{0.005,9} = 3.250$ the confidence interval is

$$\left(2.752 - \frac{3.250 \times 0.280}{\sqrt{10}}, 2.752 + \frac{3.250 \times 0.280}{\sqrt{10}}\right) = (2.464, 3.040).$$

The value 3.1 is outside this confidence interval, and so 3.1 is not a plausible value for the average corrosion rate.

Note: The sample statistics for the following problems in this section and the related problems in this chapter depend upon whether any observations have been removed as outliers. To avoid confusion, the answers given here assume that **no** observations have been removed. Notice that removing observations as outliers reduces the sample standard deviation s as well as affecting the sample mean $\bar{x}$.

8.1.19   At 95% confidence the critical point is $t_{0.025,89} = 1.987$ and the confidence interval is

$$\left(12.211 - \frac{1.987 \times 2.629}{\sqrt{90}}, 12.211 + \frac{1.987 \times 2.629}{\sqrt{90}}\right) = (11.66, 12.76).$$

8.1.21   At 95% confidence the critical point is $t_{0.025,74} = 1.9926$ and the confidence interval is

$$\left(0.23181 - \frac{1.9926 \times 0.07016}{\sqrt{75}}, 0.23181 + \frac{1.9926 \times 0.07016}{\sqrt{75}}\right) = (0.2157, 0.2480).$$

8.1.23   Since

$$2.773 = 2.843 - \frac{t_{\alpha,8} \times 0.150}{\sqrt{9}}$$

it follows that $t_{\alpha,8} = 1.40$ so that $\alpha = 0.10$.

Therefore, the confidence level of the confidence interval is 90%.

8.1.25   (a) Using the critical point $t_{0.025,13} = 2.160$ the confidence interval is

$$\mu \in 5437.2 \pm \frac{2.160 \times 376.9}{\sqrt{14}} = (5219.6, 5654.8).$$

(b) With

$$4 \times \left(\frac{2.160 \times 376.9}{300}\right)^2 = 29.5$$

it can be estimated that an additional $30 - 14 = 16$ chemical solutions would need to be measured.

8.1.27   B

8.2 Hypothesis Testing

8.2.1   (a) The test statistic is

$$t = \frac{\sqrt{n}(\bar{x} - \mu_0)}{s} = \frac{\sqrt{18} \times (57.74 - 55.0)}{11.2} = 1.04.$$

The p-value is $2 \times P(t_{17} \geq 1.04) = 0.313$.

  (b) The test statistic is

$$t = \frac{\sqrt{n}(\bar{x} - \mu_0)}{s} = \frac{\sqrt{18} \times (57.74 - 65.0)}{11.2} = -2.75.$$

The p-value is $P(t_{17} \leq -2.75) = 0.0068$.

8.2.3   (a) The test statistic is

$$z = \frac{\sqrt{n}(\bar{x} - \mu_0)}{\sigma} = \frac{\sqrt{13} \times (2.879 - 3.0)}{0.325} = -1.34.$$

The p-value is $2 \times \Phi(-1.34) = 0.180$.

  (b) The test statistic is

$$z = \frac{\sqrt{n}(\bar{x} - \mu_0)}{\sigma} = \frac{\sqrt{13} \times (2.879 - 3.1)}{0.325} = -2.45.$$

The p-value is $\Phi(-2.45) = 0.007$.

8.2.5   (a) The critical point is $t_{0.05,40} = 1.684$
and the null hypothesis is accepted when $|t| \leq 1.684$.

  (b) The critical point is $t_{0.005,40} = 2.704$
and the null hypothesis is rejected when $|t| > 2.704$.

  (c) The test statistic is

$$t = \frac{\sqrt{n}(\bar{x} - \mu_0)}{s} = \frac{\sqrt{41} \times (3.04 - 3.00)}{0.124} = 2.066.$$

The null hypothesis is rejected at size $\alpha = 0.10$
and accepted at size $\alpha = 0.01$.

(d) The p-value is $2 \times P(t_{40} \geq 2.066) = 0.045$.

8.2.7   (a) The critical point is $t_{0.05,15} = 1.753$
and the null hypothesis is accepted when $|t| \leq 1.753$.

(b) The critical point is $t_{0.005,15} = 2.947$
and the null hypothesis is rejected when $|t| > 2.947$.

(c) The test statistic is

$$t = \frac{\sqrt{n}(\bar{x} - \mu_0)}{s} = \frac{\sqrt{16} \times (1.053 - 1.025)}{0.058} = 1.931.$$

The null hypothesis is rejected at size $\alpha = 0.10$
and accepted at size $\alpha = 0.01$.

(d) The p-value is $2 \times P(t_{15} \geq 1.931) = 0.073$.

8.2.9   (a) The critical point is $t_{0.10,60} = 1.296$
and the null hypothesis is accepted when $t \leq 1.296$.

(b) The critical point is $t_{0.01,60} = 2.390$
and the null hypothesis is rejected when $t > 2.390$.

(c) The test statistic is

$$t = \frac{\sqrt{n}(\bar{x} - \mu_0)}{s} = \frac{\sqrt{61} \times (0.0768 - 0.065)}{0.0231} = 3.990.$$

The null hypothesis is rejected at size $\alpha = 0.01$
and consequently also at size $\alpha = 0.10$.

(d) The p-value is $P(t_{60} \geq 3.990) = 0.0001$.

8.2.11   Consider the hypotheses $H_0 : \mu = 44.350$ versus $H_A : \mu \neq 44.350$.
The test statistic is

$$t = \frac{\sqrt{n}(\bar{x} - \mu_0)}{s} = \frac{\sqrt{24} \times (44.364 - 44.350)}{0.019} = 3.61.$$

The p-value is $2 \times P(t_{23} \geq 3.61) = 0.0014$.

There is sufficient evidence to conclude that the machine is miscalibrated.

8.2.13   Consider the hypotheses $H_0 : \mu \leq 12.50$ versus $H_A : \mu > 12.50$.

The test statistic is

$$t = \frac{\sqrt{n}(\bar{x} - \mu_0)}{s} = \frac{\sqrt{15} \times (14.82 - 12.50)}{2.91} = 3.09.$$

The p-value is $P(t_{14} \geq 3.09) = 0.004$.

There is sufficient evidence to conclude that the chemical plant is in violation of the working code.

8.2.15   Consider the hypotheses $H_0 : \mu \leq 2.5$ versus $H_A : \mu > 2.5$.

The test statistic is

$$t = \frac{\sqrt{n}(\bar{x} - \mu_0)}{s} = \frac{\sqrt{10} \times (2.752 - 2.5)}{0.280} = 2.846.$$

The p-value is $P(t_9 \geq 2.846) = 0.0096$.

There is sufficient evidence to conclude that the average corrosion rate of chilled cast iron of this type is larger than 2.5.

8.2.17   Consider the hypotheses $H_0 : \mu \geq 13$ versus $H_A : \mu < 13$.

The test statistic is

$$t = \frac{\sqrt{n}(\bar{x} - \mu_0)}{s} = \frac{\sqrt{90} \times (12.211 - 13.000)}{2.629} = -2.85.$$

The p-value is $P(t_{89} \leq -2.85) = 0.0027$.

There is sufficient evidence to conclude that the average number of calls taken per minute is less than 13 so that the manager's claim is false.

8.2.19   Consider the hypotheses $H_0 : \mu = 0.2$ versus $H_A : \mu \neq 0.2$.

The test statistic is

$$t = \frac{\sqrt{n}(\bar{x} - \mu_0)}{s} = \frac{\sqrt{75} \times (0.23181 - 0.22500)}{0.07016} = 0.841.$$

The p-value is $2 \times P(t_{74} \geq 0.841) = 0.40$.

There is not sufficient evidence to conclude that the spray painting machine is not performing properly.

8.2.21   The hypotheses are $H_0 : \mu \leq 238.5$ versus $H_A : \mu > 238.5$

and the test statistic is

$$t = \frac{\sqrt{16}(239.13 - 238.50)}{2.80} = 0.90.$$

The p-value is $P(t_{15} > 0.90) = 0.191$.

There is not sufficient evidence to conclude that the average voltage of the batteries from the production line is at least 238.5.

8.2.23   The hypotheses are $H_0 : \mu = 82.50$ versus $H_A : \mu \neq 82.50$

and the test statistic is

$$t = \frac{\sqrt{25}(82.40 - 82.50)}{0.14} = -3.571.$$

The p-value is $2 \times P(t_{24} > 3.571) = 0.0015$.

There is sufficient evidence to conclude that the average length of the components is not 82.50.

8.2.25   The hypotheses are $H_0 : \mu = 7.000$ versus $H_A : \mu \neq 7.000$

and the test statistic is

$$t = \frac{\sqrt{28}(7.442 - 7.000)}{0.672} = 3.480.$$

The p-value is $2 \times P(t_{27} > 3.480) = 0.002$.

There is sufficient evidence to conclude that the average breaking strength is not 7.000.

8.2.27   The hypotheses are $H_0 : \mu \geq 25$ versus $H_A : \mu < 25$.

8.2.29   (a) The sample mean is $\bar{x} = 11.975$
            and the sample standard deviation is $s = 2.084$
            so that the t-statistic is

$$t = \frac{\sqrt{8}(11.975-11)}{2.084} = 1.32.$$

The p-value is $P(t_7 > 1.32)$ which is greater than 10%.

Consequently, the experiment does not provide sufficient evidence to conclude that the average time to toxicity of salmon fillets under these storage conditions is more than 11 days.

(b) With $t_{0.005,7} = 3.499$ the confidence interval is

$$11.975 \pm \frac{3.499 \times 2.084}{\sqrt{8}} = (9.40, 14.55).$$

8.2.31   D

8.2.33   A

8.2.35   B

Chapter 9

Comparing Two Population Means

9.2 Analysis of Paired Samples

9.2.1 The differences $z_i = x_i - y_i$ have a sample mean $\bar{z} = 4.257$ and a sample standard deviation $s = 37.65$.

Consider the hypotheses

$H_0 : \mu = \mu_A - \mu_B \leq 0$ versus $H_A : \mu = \mu_A - \mu_B > 0$

where the alternative hypothesis states that the new assembly method is quicker on average than the standard assembly method.

The test statistic is

$$t = \frac{\sqrt{n}\,\bar{z}}{s} = \frac{\sqrt{35} \times 4.257}{37.65} = 0.669.$$

The p-value is $P(t_{34} \geq 0.669) = 0.254$.

There is *not* sufficient evidence to conclude that the new assembly method is any quicker on average than the standard assembly method.

With $t_{0.05,34} = 1.691$ a one-sided 95% confidence level confidence interval for $\mu = \mu_A - \mu_B$ is

$$\left(4.257 - \frac{1.691 \times 37.65}{\sqrt{35}}, \infty \right) = (-6.50, \infty).$$

9.2.3 The differences $z_i = x_i - y_i$ have a sample mean $\bar{z} = 0.570$ and a sample standard deviation $s = 0.813$.

Consider the hypotheses

$H_0 : \mu = \mu_A - \mu_B \leq 0$ versus $H_A : \mu = \mu_A - \mu_B > 0$

where the alternative hypothesis states that the new tires have a smaller average reduction in tread depth than the standard tires.

The test statistic is

$t = \frac{\sqrt{n}\,\bar{z}}{s} = \frac{\sqrt{20} \times 0.570}{0.813} = 3.14.$

The p-value is $P(t_{19} \geq 3.14) = 0.003$.

There is sufficient evidence to conclude that the new tires are better than the standard tires in terms of the average reduction in tread depth.

With $t_{0.05,19} = 1.729$ a one-sided 95% confidence level confidence interval for $\mu = \mu_A - \mu_B$ is

$\left(0.570 - \frac{1.729 \times 0.813}{\sqrt{20}}, \infty \right) = (0.256, \infty).$

9.2.5 The differences $z_i = x_i - y_i$ have a sample mean $\bar{z} = 2.20$ and a sample standard deviation $s = 147.8$.

Consider the hypotheses

$H_0 : \mu = \mu_A - \mu_B = 0$ versus $H_A : \mu = \mu_A - \mu_B \neq 0.$

The test statistic is

$t = \frac{\sqrt{n}\,\bar{z}}{s} = \frac{\sqrt{18} \times 2.20}{147.8} = 0.063.$

The p-value is $2 \times P(t_{17} \geq 0.063) = 0.95$.

There is *not* sufficient evidence to conclude that the two laboratories are any different in the datings that they provide.

With $t_{0.025,17} = 2.110$ a two-sided 95% confidence level confidence interval for $\mu = \mu_A - \mu_B$ is

$\left(2.20 - \frac{2.110 \times 147.8}{\sqrt{18}}, 2.20 + \frac{2.110 \times 147.8}{\sqrt{18}} \right) = (-71.3, 75.7).$

9.2.7   The differences $z_i = x_i - y_i$ have a sample mean $\bar{z} = -2.800$ and a sample standard deviation $s = 6.215$.

The hypotheses are

$H_0 : \mu = \mu_A - \mu_B = 0$ versus $H_A : \mu = \mu_A - \mu_B \neq 0$

and the test statistic is

$t = \frac{\sqrt{10} \times (-2.800)}{6.215} = -1.425$.

The p-value is $2 \times P(t_9 \geq 1.425) = 0.188$.

There is not sufficient evidence to conclude that procedures A and B give different readings on average.

The reviewer's comments are plausible.

9.2.9   The differences $z_i = x_i - y_i$ have a sample mean $\bar{z} = 0.85$ and a sample standard deviation $s = 4.283$.

Consider the hypotheses

$H_0 : \mu = \mu_A - \mu_B = 0$ versus $H_A : \mu = \mu_A - \mu_B \neq 0$

where the alternative hypothesis states that the addition of the surfactant has an effect on the amount of uranium-oxide removed from the water.

The test statistic is

$t = \frac{\sqrt{n}\,\bar{z}}{s} = \frac{\sqrt{6} \times 0.85}{4.283} = 0.486$.

The p-value is $2 \times P(t_5 \geq 0.486) = 0.65$.

Consequently, there is *not* sufficient evidence to conclude that the addition of the surfactant has an effect on the amount of uranium-oxide removed from the water.

9.2.11   D

9.3 Analysis of Independent Samples

9.3.3   (a) The pooled variance is

$$s_p^2 = \frac{(n-1)s_x^2 + (m-1)s_y^2}{n+m-2} = \frac{((8-1)\times 44.76^2) + ((17-1)\times 38.94^2)}{8+17-2} = 1664.6.$$

With $t_{0.005,23} = 2.807$ a 99% two-sided confidence interval for $\mu_A - \mu_B$ is

$$675.1 - 702.4 \pm 2.807 \times \sqrt{1664.6} \times \sqrt{\tfrac{1}{8} + \tfrac{1}{17}}$$

$$= (-76.4, 21.8).$$

(b) Since

$$\frac{\left(\frac{44.76^2}{8} + \frac{38.94^2}{17}\right)^2}{\frac{44.76^4}{8^2 \times (8-1)} + \frac{38.94^4}{17^2 \times (17-1)}} = 12.2$$

the degrees of freedom are $\nu = 12$.

Using a critical point $t_{0.005,12} = 3.055$
a 99% two-sided confidence interval for $\mu_A - \mu_B$ is

$$675.1 - 702.4 \pm 3.055 \times \sqrt{\frac{44.76^2}{8} + \frac{38.94^2}{17}}$$

$$= (-83.6, 29.0).$$

(c) The test statistic is

$$t = \frac{\bar{x} - \bar{y}}{s_p \sqrt{\frac{1}{n} + \frac{1}{m}}} = \frac{675.1 - 702.4}{\sqrt{1664.6} \times \sqrt{\frac{1}{8} + \frac{1}{17}}} = -1.56.$$

The null hypothesis is accepted since $|t| = 1.56$ is smaller than the critical point $t_{0.005,23} = 2.807$.
The p-value is $2 \times P(t_{23} \geq 1.56) = 0.132$.

9.3.5   (a) The pooled variance is

$$s_p^2 = \frac{(n-1)s_x^2 + (m-1)s_y^2}{n+m-2} = \frac{((13-1)\times 0.00128^2) + ((15-1)\times 0.00096^2)}{13+15-2}$$

$$= 1.25 \times 10^{-6}.$$

With $t_{0.05,26} = 1.706$ a 95% one-sided confidence interval
for $\mu_A - \mu_B$ is

$$\left(-\infty, 0.0548 - 0.0569 + 1.706 \times \sqrt{1.25 \times 10^{-6}} \times \sqrt{\tfrac{1}{13} + \tfrac{1}{15}}\right)$$

$$= (-\infty, -0.0014).$$

(b) The test statistic is

$$t = \frac{\bar{x} - \bar{y}}{s_p\sqrt{\tfrac{1}{n} + \tfrac{1}{m}}} = \frac{0.0548 - 0.0569}{\sqrt{1.25 \times 10^{-6}} \times \sqrt{\tfrac{1}{13} + \tfrac{1}{15}}} = -4.95.$$

The null hypothesis is rejected at size $\alpha = 0.01$ since
$t = -4.95 < -t_{0.01,26} = -2.479$.
The null hypothesis is consequently also rejected at size $\alpha = 0.05$.
The p-value is $P(t_{26} \leq -4.95) = 0.000$.

9.3.7   (a) Since

$$\frac{\left(\frac{11.90^2}{20} + \frac{4.61^2}{25}\right)^2}{\frac{11.90^4}{20^2 \times (20-1)} + \frac{4.61^4}{25^2 \times (25-1)}} = 23.6$$

the degrees of freedom are $\nu = 23$.

Consider the hypotheses
$H_0 : \mu = \mu_A - \mu_B \geq 0$ versus $H_A : \mu = \mu_A - \mu_B < 0$
where the alternative hypothesis states that the synthetic fiber bundles have an average breaking strength larger than the wool fiber bundles.

The test statistic is

$$t = \frac{\bar{x} - \bar{y}}{\sqrt{\tfrac{s_x^2}{n} + \tfrac{s_y^2}{m}}} = \frac{436.5 - 452.8}{\sqrt{\tfrac{11.90^2}{20} + \tfrac{4.61^2}{25}}} = -5.788.$$

The null hypothesis is rejected at size $\alpha = 0.01$ since
$t = -5.788 < -t_{0.01,23} = -2.500$.
The p-value is $P(t_{23} \leq -5.788) = 0.000$.

(b) With a critical point $t_{0.01,23} = 2.500$
a 99% one-sided confidence interval for $\mu_A - \mu_B$ is

$$\left(-\infty, 436.5 - 452.8 + 2.500 \times \sqrt{\tfrac{11.90^2}{20} + \tfrac{4.61^2}{25}} \right)$$

$$= (-\infty, -9.3).$$

(c) There is sufficient evidence to conclude that the synthetic fiber bundles have an average breaking strength larger than the wool fiber bundles.

9.3.9 (a) The test statistic is

$$z = \frac{\bar{x} - \bar{y} - \delta}{\sqrt{\frac{\sigma_A^2}{n} + \frac{\sigma_B^2}{m}}} = \frac{100.85 - 89.32 - 3}{\sqrt{\frac{25^2}{47} + \frac{20^2}{62}}} = 1.92$$

and the p-value is $2 \times \Phi(-1.92) = 0.055$.

(b) With a critical point $z_{0.05} = 1.645$ a 90% two-sided confidence interval for $\mu_A - \mu_B$ is

$$100.85 - 89.32 \pm 1.645 \times \sqrt{\tfrac{25^2}{47} + \tfrac{20^2}{62}}$$

$$= (4.22, 18.84).$$

9.3.11 (a) The test statistic is

$$z = \frac{\bar{x} - \bar{y}}{\sqrt{\frac{\sigma_A^2}{n} + \frac{\sigma_B^2}{m}}} = \frac{19.50 - 18.64}{\sqrt{\frac{1.0^2}{10} + \frac{1.0^2}{12}}} = 2.009$$

and the p-value is $2 \times \Phi(-2.009) = 0.045$.

(b) With a critical point $z_{0.05} = 1.645$ a 90% two-sided confidence interval for $\mu_A - \mu_B$ is

$$19.50 - 18.64 \pm 1.645 \times \sqrt{\tfrac{1.0^2}{10} + \tfrac{1.0^2}{12}}$$

$$= (0.16, 1.56).$$

With a critical point $z_{0.025} = 1.960$ a 95% two-sided confidence interval for $\mu_A - \mu_B$ is

$$19.50 - 18.64 \pm 1.960 \times \sqrt{\tfrac{1.0^2}{10} + \tfrac{1.0^2}{12}}$$

$$= (0.02, 1.70).$$

With a critical point $z_{0.005} = 2.576$ a 99% two-sided confidence interval for $\mu_A - \mu_B$ is

$$19.50 - 18.64 \pm 2.576 \times \sqrt{\frac{1.0^2}{10} + \frac{1.0^2}{12}}$$

$$= (-0.24, 1.96).$$

9.3.13   Using 2.0 as an upper bound for $t_{0.025, \nu}$ equal sample sizes of

$$n = m \geq \frac{4\, t_{\alpha/2, \nu}^2\, (\sigma_A^2 + \sigma_B^2)}{L_0^2} = \frac{4 \times 2.0^2 \times (1.2^2 + 1.2^2)}{1.0^2} = 46.08$$

should be sufficient.

Equal sample sizes of at least 47 can be recommended.

9.3.15   Using $t_{0.005, 80} = 2.639$ equal total sample sizes of

$$n = m \geq \frac{4\, t_{\alpha/2, \nu}^2\, (s_x^2 + s_y^2)}{L_0^2} = \frac{4 \times 2.639^2 \times (0.124^2 + 0.137^2)}{0.1^2} = 95.1$$

should be sufficient.

Additional sample sizes of at least $96 - 41 = 55$ from each population can be recommended.

9.3.17   There is sufficient evidence to conclude that the paving slabs from company A weigh more on average than the paving slabs from company B.

There is also more variability in the weights of the paving slabs from company A.

9.3.19   There is sufficient evidence to conclude that the damped feature is effective in reducing the heel-strike force.

9.3.21   There is not sufficient evidence to conclude that the average service times are any different at these two times of day.

9.3.23 $\bar{x}_A = 142.4$

$s_A = 9.24$

$n_A = 10$

$\bar{x}_B = 131.6$

$s_B = 7.97$

$n_B = 10$

The hypotheses are

$H_0 : \mu_A \leq \mu_B$ versus $H_A : \mu_A > \mu_B$

and

$$\frac{\left(\frac{9.24^2}{10} + \frac{7.97^2}{10}\right)^2}{\frac{9.24^4}{10^2 \times (10-1)} + \frac{7.97^4}{10^2 \times (10-1)}} = 17.6$$

so that the degrees of freedom are $\nu = 17$.

The test statistic is

$$t = \frac{142.4 - 131.6}{\sqrt{\frac{9.24^2}{10} + \frac{7.97^2}{10}}} = 2.799$$

and the p-value is $P(t_{17} > 2.799) = 0.006$.

There is sufficient evidence to conclude that on average medicine A provides a higher response than medicine B.

9.3.25 $\bar{x}_A = 152.3$

$s_A = 1.83$

$n_A = 10$

$s_B = 1.94$

$n_B = 8$

The hypotheses are

$H_0 : \mu_A \leq \mu_B$ versus $H_A : \mu_A > \mu_B$

and

$$\frac{\left(\frac{1.83^2}{10} + \frac{1.94^2}{8}\right)^2}{\frac{1.83^4}{10^2 \times (10-1)} + \frac{1.94^4}{8^2 \times (8-1)}} = 14.7$$

so that the degrees of freedom are $\nu = 14$.

Since the p-value is $P(t_{14} > t) < 0.01$, it follows that

$$t = \frac{\bar{x}_A - \bar{x}_B}{\sqrt{\frac{s_A^2}{n_A} + \frac{s_B^2}{n_B}}} = \frac{152.3 - \bar{x}_B}{0.8974} > t_{0.01,14} = 2.624$$

so that $\bar{x}_B < 149.9$.

Chapter 10

Discrete Data Analysis

10.1 Inferences on a Population Proportion

10.1.1   (a) With $z_{0.005} = 2.576$ the confidence interval is

$$\left(\frac{11}{32} - \frac{2.576}{32} \times \sqrt{\frac{11 \times (32-11)}{32}}, \frac{11}{32} + \frac{2.576}{32} \times \sqrt{\frac{11 \times (32-11)}{32}} \right)$$

$$= (0.127, 0.560).$$

(b) With $z_{0.025} = 1.960$ the confidence interval is

$$\left(\frac{11}{32} - \frac{1.960}{32} \times \sqrt{\frac{11 \times (32-11)}{32}}, \frac{11}{32} + \frac{1.960}{32} \times \sqrt{\frac{11 \times (32-11)}{32}} \right)$$

$$= (0.179, 0.508).$$

(c) With $z_{0.01} = 2.326$ the confidence interval is

$$\left(0, \frac{11}{32} + \frac{2.326}{32} \times \sqrt{\frac{11 \times (32-11)}{32}} \right)$$

$$= (0, 0.539).$$

(d) The exact p-value is $2 \times P(B(32, 0.5) \leq 11) = 0.110$.

The statistic for the normal approximation to the p-value is

$$z = \frac{x - np_0}{\sqrt{np_0(1-p_0)}} = \frac{11 - (32 \times 0.5)}{\sqrt{32 \times 0.5 \times (1-0.5)}} = -1.768$$

and the p-value is $2 \times \Phi(-1.768) = 0.077$.

113

10.1.3   (a) Let p be the probability that a value produced by the random number generator is a zero, and consider the hypotheses

$H_0 : p = 0.5$ versus $H_A : p \neq 0.5$

where the alternative hypothesis states that the random number generator is producing 0's and 1's with unequal probabilities.

The statistic for the normal approximation to the p-value is

$$z = \frac{x - np_0}{\sqrt{np_0(1-p_0)}} = \frac{25264 - (50000 \times 0.5)}{\sqrt{50000 \times 0.5 \times (1-0.5)}} = 2.361$$

and the p-value is $2 \times \Phi(-2.361) = 0.018$.

There is a fairly strong suggestion that the random number generator is producing 0's and 1's with unequal probabilities, although the evidence is not completely overwhelming.

(b) With $z_{0.005} = 2.576$ the confidence interval is

$$\left(\frac{25264}{50000} - \frac{2.576}{50000} \times \sqrt{\frac{25264 \times (50000 - 25264)}{50000}}, \frac{25264}{50000} + \frac{2.576}{50000} \times \sqrt{\frac{25264 \times (50000 - 25264)}{50000}} \right)$$

$$= (0.4995, 0.5110).$$

(c) Using the worst case scenario

$\hat{p}(1 - \hat{p}) = 0.25$

the total sample size required can be calculated as

$$n \geq \frac{4 \, z_{\alpha/2}^2 \, \hat{p}(1-\hat{p})}{L^2}$$

$$= \frac{4 \times 2.576^2 \times 0.25}{0.005^2} = 265431.04$$

so that an additional sample size of $265432 - 50000 \simeq 215500$ would be required.

10.1.5   Let p be the probability that a six is scored on the die and consider the hypotheses

$$H_0 : p \geq \tfrac{1}{6} \text{ versus } H_A : p < \tfrac{1}{6}$$

where the alternative hypothesis states that the die has been weighted to reduce the chance of scoring a six.

In the first experiment the exact p-value is

$$P\left(B\left(50, \tfrac{1}{6}\right) \leq 2\right) = 0.0066$$

and in the second experiment the exact p-value is

$$P\left(B\left(100, \tfrac{1}{6}\right) \leq 4\right) = 0.0001$$

so that there is more support for foul play from the second experiment than from the first.

10.1.7   Let p be the probability that a juror is selected from the county where the investigator lives, and consider the hypotheses

$$H_0 : p = 0.14 \text{ versus } H_A : p \neq 0.14$$

where the alternative hypothesis implies that the jurors are not being randomly selected.

The statistic for the normal approximation to the p-value is

$$z = \frac{x - np_0}{\sqrt{np_0(1-p_0)}} = \frac{122 - (1{,}386 \times 0.14)}{\sqrt{1{,}386 \times 0.14 \times (1 - 0.14)}} = -5.577$$

and the p-value is $2 \times \Phi(-5.577) = 0.000$.

There is sufficient evidence to conclude that the jurors are not being randomly selected.

10.1.9   With $z_{0.025} = 1.960$ and $L = 0.02$

the required sample size for the worst case scenario with

$$\hat{p}(1 - \hat{p}) = 0.25$$

can be calculated as

$$n \geq \frac{4\, z_{\alpha/2}^2\, \hat{p}(1-\hat{p})}{L^2} = \frac{4 \times 1.960^2 \times 0.25}{0.02^2} = 9604.$$

If it can be assumed that

$$\hat{p}(1 - \hat{p}) \leq 0.75 \times 0.25 = 0.1875$$

then the required sample size can be calculated as

$$n \geq \frac{4\, z_{\alpha/2}^2\, \hat{p}(1-\hat{p})}{L^2} = \frac{4 \times 1.960^2 \times 0.1875}{0.02^2} = 7203.$$

10.1.11   With $z_{0.005} = 2.576$ the confidence interval is

$$\left(\frac{73}{120} - \frac{2.576}{120} \times \sqrt{\frac{73 \times (120-73)}{120}}, \frac{73}{120} + \frac{2.576}{120} \times \sqrt{\frac{73 \times (120-73)}{120}} \right)$$

$$= (0.494, 0.723).$$

Using

$$\hat{p}(1 - \hat{p}) = \frac{73}{120} \times \left(1 - \frac{73}{120} \right) = 0.238$$

the total sample size required can be calculated as

$$n \geq \frac{4\, z_{\alpha/2}^2\, \hat{p}(1-\hat{p})}{L^2} = \frac{4 \times 2.576^2 \times 0.238}{0.1^2} = 631.7$$

so that an additional sample size of $632 - 120 = 512$ would be required.

10.1.13   With $z_{0.025} = 1.960$ the confidence interval is

$$\left(\frac{12}{20} - \frac{1.960}{20} \times \sqrt{\frac{12 \times (20-12)}{20}}, \frac{12}{20} + \frac{1.960}{20} \times \sqrt{\frac{12 \times (20-12)}{20}} \right)$$

$$= (0.385, 0.815).$$

10.1.15   With $z_{0.025} = 1.960$ and $L = 0.10$

the required sample size for the worst case scenario with

$\hat{p}(1 - \hat{p}) = 0.25$

can be calculated as

$$n \geq \frac{4\, z_{\alpha/2}^2\, \hat{p}(1-\hat{p})}{L^2} = \frac{4 \times 1.960^2 \times 0.25}{0.10^2} = 384.2$$

or 385 householders.

If it can be assumed that

$\hat{p}(1 - \hat{p}) \leq 0.333 \times 0.667 = 0.222$

then the required sample size can be calculated as

$$n \geq \frac{4\, z_{\alpha/2}^2\, \hat{p}(1-\hat{p})}{L^2} = \frac{4 \times 1.960^2 \times 0.222}{0.10^2} = 341.1$$

or 342 householders.

10.1.17   The standard confidence interval is $(0.161, 0.557)$.

The alternative confidence interval is $(0.195, 0.564)$.

10.1.19 $\hat{p} = \frac{31}{210} = 0.148$

With $z_{0.005} = 2.576$ the confidence interval is

$p \in 0.148 \pm \frac{2.576}{210} \sqrt{\frac{31 \times (210-31)}{210}}$

$= (0.085, 0.211)$.

10.1.21   If $793 = \frac{z_{\alpha/2}^2}{(2 \times 0.035)^2}$

then $z_{\alpha/2}^2 = 1.97$ so that $\alpha \simeq 0.05$.

Therefore, the margin of error was calculated with 95% confidence under the worst case scenario where the estimated probability could be close to 0.5.

10.2   Comparing Two Population Proportions

10.2.1   (a) With $z_{0.005} = 2.576$ the confidence interval is

$$\frac{14}{37} - \frac{7}{26} \pm 2.576 \times \sqrt{\frac{14 \times (37-14)}{37^3} + \frac{7 \times (26-7)}{26^3}}$$

$$= (-0.195, 0.413).$$

(b) With $z_{0.025} = 1.960$ the confidence interval is

$$\frac{14}{37} - \frac{7}{26} \pm 1.960 \times \sqrt{\frac{14 \times (37-14)}{37^3} + \frac{7 \times (26-7)}{26^3}}$$

$$= (-0.122, 0.340).$$

(c) With $z_{0.01} = 2.326$ the confidence interval is

$$\left(\frac{14}{37} - \frac{7}{26} - 2.326 \times \sqrt{\frac{14 \times (37-14)}{37^3} + \frac{7 \times (26-7)}{26^3}}, 1 \right)$$

$$= (-0.165, 1).$$

(d) With the pooled probability estimate

$$\hat{p} = \frac{x+y}{n+m} = \frac{14+7}{37+26} = 0.333$$

the test statistic is

$$z = \frac{\hat{p}_A - \hat{p}_B}{\sqrt{\hat{p}(1-\hat{p})\left(\frac{1}{n}+\frac{1}{m}\right)}} = \frac{\frac{14}{37} - \frac{7}{26}}{\sqrt{0.333 \times (1-0.333) \times \left(\frac{1}{37}+\frac{1}{26}\right)}} = 0.905$$

and the p-value is $2 \times \Phi(-0.905) = 0.365$.

10.2.3   (a) With $z_{0.005} = 2.576$ the confidence interval is

$$\frac{35}{44} - \frac{36}{52} \pm 2.576 \times \sqrt{\frac{35 \times (44-35)}{44^3} + \frac{36 \times (52-36)}{52^3}}$$

$$= (-0.124, 0.331).$$

(b) With the pooled probability estimate

$$\hat{p} = \frac{x+y}{n+m} = \frac{35+36}{44+52} = 0.740$$

the test statistic is

$$z = \frac{\hat{p}_A - \hat{p}_B}{\sqrt{\hat{p}(1-\hat{p})\left(\frac{1}{n}+\frac{1}{m}\right)}} = \frac{\frac{35}{44} - \frac{36}{52}}{\sqrt{0.740 \times (1-0.740) \times \left(\frac{1}{44}+\frac{1}{52}\right)}} = 1.147$$

and the p-value is $2 \times \Phi(-1.147) = 0.251$.

There is *not* sufficient evidence to conclude that one radar system is any better than the other radar system.

10.2.5   Let p_A be the probability of crystallization within 24 hours *without* seed crystals and let p_B be the probability of crystallization within 24 hours *with* seed crystals.

With $z_{0.05} = 1.645$ a 95% upper confidence bound for $p_A - p_B$ is

$$\left(-1, \frac{27}{60} - \frac{36}{60} + 1.645 \times \sqrt{\frac{27 \times (60-27)}{60^3} + \frac{36 \times (60-36)}{60^3}}\right)$$

$$= (-1, -0.002).$$

Consider the hypotheses

$H_0 : p_A \geq p_B$ versus $H_A : p_A < p_B$

where the alternative hypothesis states that the presence of seed crystals increases the probability of crystallization within 24 hours.

With the pooled probability estimate

$\hat{p} = \frac{x+y}{n+m} = \frac{27+36}{60+60} = 0.525$

the test statistic is

$$z = \frac{\hat{p}_A - \hat{p}_B}{\sqrt{\hat{p}(1-\hat{p})\left(\frac{1}{n}+\frac{1}{m}\right)}} = \frac{\frac{27}{60} - \frac{36}{60}}{\sqrt{0.525 \times (1-0.525) \times \left(\frac{1}{60}+\frac{1}{60}\right)}} = -1.645$$

and the p-value is $\Phi(-1.645) = 0.050$.

There is some evidence that the presence of seed crystals increases the probability of crystallization within 24 hours but it is not overwhelming.

10.2.7   Let p_A be the probability that a television set from production line A does not meet the quality standards and let p_B be the probability that a television set from production line B does not meet the quality standards.

With $z_{0.025} = 1.960$ a 95% two-sided confidence interval for $p_A - p_B$ is

$$\frac{23}{1128} - \frac{24}{962} \pm 1.960 \times \sqrt{\frac{23 \times (1128-23)}{1128^3} + \frac{24 \times (962-24)}{962^3}}$$

$$= (-0.017, 0.008).$$

Consider the hypotheses

$H_0 : p_A = p_B$ versus $H_A : p_A \neq p_B$

where the alternative hypothesis states that there is a difference in the operating standards of the two production lines.

With the pooled probability estimate

$$\hat{p} = \frac{x+y}{n+m} = \frac{23+24}{1128+962} = 0.022$$

the test statistic is

$$z = \frac{\hat{p}_A - \hat{p}_B}{\sqrt{\hat{p}(1-\hat{p})\left(\frac{1}{n}+\frac{1}{m}\right)}} = \frac{\frac{23}{1128} - \frac{24}{962}}{\sqrt{0.022 \times (1-0.022) \times \left(\frac{1}{1128}+\frac{1}{962}\right)}} = -0.708$$

and the p-value is $2 \times \Phi(-0.708) = 0.479$.

There is *not* sufficient evidence to conclude that there is a difference in the operating standards of the two production lines.

10.2.9   Let p_A be the probability that a computer chip from supplier A is defective and let p_B be the probability that a computer chip from supplier B is defective.

With $z_{0.025} = 1.960$ a 95% two-sided confidence interval for $p_A - p_B$ is

$$\frac{8}{200} - \frac{13}{250} \pm 1.960 \times \sqrt{\frac{8 \times (200-8)}{200^3} + \frac{13 \times (250-13)}{250^3}}$$

$$= (-0.051, 0.027).$$

Consider the hypotheses

$H_0 : p_A = p_B$ versus $H_A : p_A \neq p_B$

where the alternative hypothesis states that there is a difference in the quality of the computer chips from the two suppliers.

With the pooled probability estimate

$$\hat{p} = \frac{x+y}{n+m} = \frac{8+13}{200+250} = 0.047$$

the test statistic is

$$z = \frac{\hat{p}_A - \hat{p}_B}{\sqrt{\hat{p}(1-\hat{p})\left(\frac{1}{n}+\frac{1}{m}\right)}} = \frac{\frac{8}{200} - \frac{13}{250}}{\sqrt{0.047 \times (1-0.047) \times \left(\frac{1}{200}+\frac{1}{250}\right)}} = -0.600$$

and the p-value is $2 \times \Phi(-0.600) = 0.549$.

There is *not* sufficient evidence to conclude that there is a difference in the quality of the computer chips from the two suppliers.

10.2.11   With the pooled probability estimate

$$\hat{p} = \frac{x+y}{n+m} = \frac{159+138}{185+185} = 0.803$$

the test statistic is

$$z = \frac{\hat{p}_A - \hat{p}_B}{\sqrt{\hat{p}(1-\hat{p})\left(\frac{1}{n}+\frac{1}{m}\right)}} = \frac{\frac{159}{185} - \frac{138}{185}}{\sqrt{0.803 \times (1-0.803) \times \left(\frac{1}{185}+\frac{1}{185}\right)}} = 2.745$$

and the two-sided p-value is $2 \times \Phi(-2.745) = 0.006$.

The two-sided null hypothesis $H_0 : p_A = p_B$ is rejected and there is sufficient evidence to conclude that machine A is better than machine B.

10.2.13   (a)   Consider the hypotheses

$H_0 : p_{180} \geq p_{250}$ versus $H_A : p_{180} < p_{250}$

where the alternative hypothesis states that the probability of an insulator of this type having a dielectric breakdown strength below the specified threshold level is larger at 250 degrees Centigrade than it is at 180 degrees Centigrade.

With the pooled probability estimate

$$\frac{x+y}{n+m} = \frac{13+20}{62+70} = 0.25$$

the test statistic is

$$z = \frac{\hat{p}_{180} - \hat{p}_{250}}{\sqrt{\hat{p}(1-\hat{p})\left(\frac{1}{n}+\frac{1}{m}\right)}} = \frac{\frac{13}{62} - \frac{20}{70}}{\sqrt{0.25 \times (1-0.25)\left(\frac{1}{62}+\frac{1}{70}\right)}} = -1.007$$

and the p-value is $\Phi(-1.007) = 0.1570$.

There is not sufficient evidence to conclude that the probability of an insulator of this type having a dielectric breakdown strength below the specified threshold level is larger at 250 degrees Centigrade than it is at 180 degrees Centigrade.

(b) With $z_{0.005} = 2.576$ the confidence interval is

$$\frac{13}{62} - \frac{20}{70} \pm 2.576 \times \sqrt{\frac{13 \times (62-13)}{62^3} + \frac{20 \times (70-20)}{70^3}}$$

$$= (-0.269, 0.117).$$

10.2.15 $\hat{p}_1 = \frac{76}{243} = 0.313$

$\hat{p}_2 = \frac{122}{320} = 0.381$

With $z_{0.005} = 2.576$ the confidence interval is

$$p_1 - p_2 \in 0.313 - 0.381 \pm 2.576 \times \sqrt{\frac{76 \times (243-76)}{243^3} + \frac{122 \times (320-122)}{320^3}}$$

$$= (-0.172, 0.036)$$

The confidence interval contains zero so there is not sufficient evidence to conclude that the failure rates due to operator misuse are different for the two products.

10.3 One-way Contingency Tables

10.3.1 (a) The expected cell frequencies are $e_i = \frac{500}{6} = 83.33$.

(b) The Pearson chi-square statistic is

$$X^2 = \frac{(80-83.33)^2}{83.33} + \frac{(71-83.33)^2}{83.33} + \frac{(90-83.33)^2}{83.33} + \frac{(87-83.33)^2}{83.33}$$
$$+ \frac{(78-83.33)^2}{83.33} + \frac{(94-83.33)^2}{83.33} = 4.36.$$

(c) The likelihood ratio chi-square statistic is

$$G^2 = 2 \times \left(80 \ln \left(\frac{80}{83.33} \right) + 71 \ln \left(\frac{71}{83.33} \right) + 90 \ln \left(\frac{90}{83.33} \right) \right.$$
$$\left. + 87 \ln \left(\frac{87}{83.33} \right) + 78 \ln \left(\frac{78}{83.33} \right) + 94 \ln \left(\frac{94}{83.33} \right) \right) = 4.44.$$

(d) The p-values are $P(\chi_5^2 \geq 4.36) = 0.499$ and $P(\chi_5^2 \geq 4.44) = 0.488$.

A size $\alpha = 0.01$ test of the null hypothesis that the die is fair is accepted.

(e) With $z_{0.05} = 1.645$ the confidence interval is

$$\left(\frac{94}{500} - \frac{1.645}{500} \times \sqrt{\frac{94 \times (500-94)}{500}}, \frac{94}{500} + \frac{1.645}{500} \times \sqrt{\frac{94 \times (500-94)}{500}} \right)$$
$$= (0.159, 0.217).$$

10.3.3 (a) The expected cell frequencies are:

$e_1 = 221 \times \frac{4}{7} = 126.29$

$e_2 = 221 \times \frac{2}{7} = 63.14$

$e_3 = 221 \times \frac{1}{7} = 31.57$

The Pearson chi-square statistic is

$$X^2 = \frac{(113-126.29)^2}{126.29} + \frac{(82-63.14)^2}{63.14} + \frac{(26-31.57)^2}{31.57} = 8.01.$$
The p-value is $P(\chi_2^2 \geq 8.01) = 0.018$.

There is a fairly strong suggestion that the supposition is not plausible although the evidence is not completely overwhelming.

(b) With $z_{0.005} = 2.576$ the confidence interval is

$$\left(\frac{113}{221} - \frac{2.576}{221} \times \sqrt{\frac{113 \times (221 - 113)}{221}}, \frac{113}{221} + \frac{2.576}{221} \times \sqrt{\frac{113 \times (221 - 113)}{221}} \right)$$

$$= (0.425, 0.598).$$

10.3.5   (a) The expected cell frequencies are:

$e_1 = 126 \times 0.5 = 63.0$

$e_2 = 126 \times 0.4 = 50.4$

$e_3 = 126 \times 0.1 = 12.6$

The likelihood ratio chi-square statistic is

$$G^2 = 2 \times \left(56 \ln \left(\frac{56}{63.0} \right) + 51 \ln \left(\frac{51}{50.4} \right) + 19 \ln \left(\frac{19}{12.6} \right) \right) = 3.62.$$

The p-value is $P(\chi_2^2 \geq 3.62) = 0.164$.

These probability values are plausible.

(b) With $z_{0.025} = 1.960$ the confidence interval is

$$\left(\frac{56}{126} - \frac{1.960}{126} \times \sqrt{\frac{56 \times (126 - 56)}{126}}, \frac{56}{126} + \frac{1.960}{126} \times \sqrt{\frac{56 \times (126 - 56)}{126}} \right)$$

$$= (0.358, 0.531).$$

10.3.7   The first two cells should be pooled so that there are 13 cells altogether.

The Pearson chi-square statistic is $X^2 = 92.9$
and the p-value is $P(\chi_{12}^2 \geq 92.9) = 0.0000$.

It is not reasonable to model the number of arrivals with a
Poisson distribution with mean $\lambda = 7$.

10.3.9   According to the genetic theory the probabilities are $\frac{9}{16}$, $\frac{3}{16}$, $\frac{3}{16}$ and $\frac{1}{16}$,
so that the expected cell frequencies are:

$$e_1 = \frac{9 \times 727}{16} = 408.9375$$

$$e_2 = \frac{3 \times 727}{16} = 136.3125$$

$$e_3 = \frac{3 \times 727}{16} = 136.3125$$

$$e_4 = \frac{1 \times 727}{16} = 45.4375$$

The Pearson chi-square statistic is

$$X^2 = \frac{(412-408.9375)^2}{408.9375} + \frac{(121-136.3125)^2}{136.3125}$$

$$+ \frac{(148-136.3125)^2}{136.3125} + \frac{(46-45.4375)^2}{45.4375} = 2.75$$

and the likelihood ratio chi-square statistic is

$$G^2 = 2 \times \left(412 \ln\left(\frac{412}{408.9375}\right) + 121 \ln\left(\frac{121}{136.3125}\right) \right.$$

$$\left. + 148 \ln\left(\frac{148}{136.3125}\right) + 46 \ln\left(\frac{46}{45.4375}\right) \right) = 2.79.$$

The p-values are $P(\chi_3^2 \geq 2.75) = 0.432$ and $P(\chi_3^2 \geq 2.79) = 0.425$

so that the data set is consistent with the proposed genetic theory.

10.3.11   (a) $\hat{p}_3 = \frac{489}{630} = 0.776$

       The hypotheses are
       $H_0 : p_3 = 0.80$ versus $H_A : p_3 \neq 0.80$
       and the test statistic is

$$z = \frac{489 - (630 \times 0.8)}{\sqrt{630 \times 0.8 \times 0.2}} = -1.494.$$

       The p-value is $2 \times \Phi(-1.494) = 0.135$.

       There is not sufficient evidence to conclude that the probability that a solution has normal acidity is not 0.80.

   (b) $e_1 = 630 \times 0.04 = 25.2$
         $e_2 = 630 \times 0.06 = 37.8$
         $e_3 = 630 \times 0.80 = 504.0$
         $e_4 = 630 \times 0.06 = 37.8$

$$e_5 = 630 \times 0.04 = 25.2$$

The Pearson chi-square statistic is

$$X^2 = \frac{(34-25.2)^2}{25.2} + \frac{(41-37.8)^2}{37.8} + \frac{(489-504.0)^2}{504.0} + \frac{(52-37.8)^2}{37.8} + \frac{(14-25.2)^2}{25.2}$$
$$= 14.1$$

so that the p-value is $P(X_4^2 \geq 14.1) = 0.007$.

The data are not consistent with the claimed probabilities.

10.3.13   The total sample size is $n = 76$.

Under the specified Poisson distribution the expected cell frequencies are:

$$e_1 = 76 \times e^{-2.5} \times \frac{2.5^0}{0!} = 6.238$$

$$e_2 = 76 \times e^{-2.5} \times \frac{2.5^1}{1!} = 15.596$$

$$e_3 = 76 \times e^{-2.5} \times \frac{2.5^2}{2!} = 19.495$$

$$e_4 = 76 \times e^{-2.5} \times \frac{2.5^3}{3!} = 16.246$$

$$e_5 = 76 \times e^{-2.5} \times \frac{2.5^4}{4!} = 10.154$$

$$e_6 = 76 - e_1 - e_2 - e_3 - e_4 - e_5 = 8.270$$

The Pearson chi-square statistic is

$$X^2 = \frac{(3-6.238)^2}{6.238} + \frac{(12-15.596)^2}{15.596} + \frac{(23-19.495)^2}{19.495}$$

$$+ \frac{(18-16.246)^2}{16.246} + \frac{(13-10.154)^2}{10.154} + \frac{(7-8.270)^2}{8.270} = 4.32$$

so that the p-value is $P(\chi_5^2 \geq 4.32) = 0.50$.

It is plausible that the number of shark attacks per year follows a Poisson distribution with mean 2.5.

10.3.15   D

10.4 Two-way Contingency Tables

10.4.1   (a) The expected cell frequencies are

	Acceptable	Defective
Supplier A	186.25	13.75
Supplier B	186.25	13.75
Supplier C	186.25	13.75
Supplier D	186.25	13.75

(b) The Pearson chi-square statistic is $X^2 = 7.087$.

(c) The likelihood ratio chi-square statistic is $G^2 = 6.889$.

(d) The p-values are $P(\chi_3^2 \geq 7.087) = 0.069$ and $P(\chi_3^2 \geq 6.889) = 0.076$ where the degrees of freedom of the chi-square random variable are calculated as $(4-1) \times (2-1) = 3$.

(e) The null hypothesis that the defective rates are identical for the four suppliers is accepted at size $\alpha = 0.05$.

(f) With $z_{0.025} = 1.960$ the confidence interval is

$$\frac{10}{200} \pm \frac{1.960}{200} \times \sqrt{\frac{10 \times (200-10)}{200}}$$
$$= (0.020, 0.080).$$

(g) With $z_{0.025} = 1.960$ the confidence interval is

$$\frac{15}{200} - \frac{21}{200} \pm 1.960 \times \sqrt{\frac{15 \times (200-15)}{200^3} + \frac{21 \times (200-21)}{200^3}}$$
$$= (-0.086, 0.026).$$

10.4.3   The expected cell frequencies are

	Formulation I	Formulation II	Formulation III
10-25	75.00	74.33	50.67
26-50	75.00	74.33	50.67
≥ 51	75.00	74.33	50.67

The Pearson chi-square statistic is $X^2 = 6.11$.

The p-value is $P(\chi^2_4 \geq 6.11) = 0.191$ where the degrees of freedom of the chi-square random variable are calculated as $(3-1) \times (3-1) = 4$.

There is *not* sufficient evidence to conclude that the preferences for the different formulations change with age.

10.4.5   The expected cell frequencies are

	Completely satisfied	Somewhat satisfied	Not satisfied
Technician 1	71.50	22.36	4.14
Technician 2	83.90	26.24	4.86
Technician 3	45.96	14.37	2.66
Technician 4	57.64	18.03	3.34

The Pearson chi-square statistic is $X^2 = 32.11$.

The p-value is $P(\chi^2_6 \geq 32.11) = 0.000$ where the degrees of freedom of the chi-square random variable are calculated as $(4-1) \times (3-1) = 6$.

There is sufficient evidence to conclude that some technicians are better than others in satisfying their customers.

Note: In this analysis 4 of the cells have expected values less than 5 and it may be preferable to pool together the categories "somewhat satisfied" and "not satisfied". In this case the Pearson chi-square statistic is $X^2 = 31.07$ and comparison with a chi-square distribution with 3 degrees of freedom again gives a p-value of 0.000. The conclusion remains the same.

10.4.7   (a)   The expected cell frequencies are

	Less than one week	More than one week
Standard drug	88.63	64.37
New drug	79.37	57.63

The Pearson chi-square statistic is $X^2 = 15.71$.

The p-value is $P(\chi_1^2 \geq 15.71) = 0.0000$ where the degrees of freedom of the chi-square random variable are calculated as $(2-1) \times (2-1) = 1$.

There is sufficient evidence to conclude that $p_s \neq p_n$.

(b)   With $z_{0.005} = 2.576$ the confidence interval is

$$\frac{72}{153} - \frac{96}{137} \pm 2.576 \times \sqrt{\frac{72 \times (153-72)}{153^3} + \frac{96 \times (137-96)}{137^3}}$$
$$= (-0.375, -0.085).$$

10.4.9   The expected cell frequencies are

Type	Warranty purchased	Warranty not purchased
A	34.84	54.16
B	58.71	91.29
C	43.45	67.55

The Pearson chi-square statistic is $X^2 = 2.347$.

The p-value is $P(\chi_2^2 \geq 2.347) = 0.309$.

The null hypothesis of independence is plausible and there is not sufficient evidence to conclude that the probability of a customer purchasing the extended warranty is different for the three product types.

10.4.11   A

Chapter 11

The Analysis of Variance

11.1 One Factor Analysis of Variance

11.1.1   (a) $P(X \geq 4.2) = 0.0177$

       (b) $P(X \geq 2.3) = 0.0530$

       (c) $P(X \geq 31.7) \leq 0.0001$

       (d) $P(X \geq 9.3) = 0.0019$

       (e) $P(X \geq 0.9) = 0.5010$

11.1.3

Source	df	SS	MS	F	p-value
Treatments	7	126.95	18.136	5.01	0.0016
Error	22	79.64	3.62		
Total	29	206.59			

11.1.5

Source	df	SS	MS	F	p-value
Treatments	3	162.19	54.06	6.69	0.001
Error	40	323.34	8.08		
Total	43	485.53			

131

11.1.7

Source	df	SS	MS	F	p-value
Treatments	3	0.0079	0.0026	1.65	0.189
Error	52	0.0829	0.0016		
Total	55	0.0908			

11.1.9   (a) $\mu_1 - \mu_2 \in \left(136.3 - 152.1 - \frac{\sqrt{15.95} \times 4.30}{\sqrt{6}}, 136.3 - 152.1 + \frac{\sqrt{15.95} \times 4.30}{\sqrt{6}}\right)$

$= (-22.8, -8.8)$

$\mu_1 - \mu_3 \in (3.6, 17.6)$

$\mu_1 - \mu_4 \in (-0.9, 13.1)$

$\mu_1 - \mu_5 \in (-13.0, 1.0)$

$\mu_1 - \mu_6 \in (1.3, 15.3)$

$\mu_2 - \mu_3 \in (19.4, 33.4)$

$\mu_2 - \mu_4 \in (14.9, 28.9)$

$\mu_2 - \mu_5 \in (2.8, 16.8)$

$\mu_2 - \mu_6 \in (17.1, 31.1)$

$\mu_3 - \mu_4 \in (-11.5, 2.5)$

$\mu_3 - \mu_5 \in (-23.6, -9.6)$

$\mu_3 - \mu_6 \in (-9.3, 4.7)$

$\mu_4 - \mu_5 \in (-19.1, -5.1)$

$\mu_4 - \mu_6 \in (-4.8, 9.2)$

$\mu_5 - \mu_6 \in (7.3, 21.3)$

(c) The total sample size required from each factor level can be estimated as

$$n \geq \frac{4\, s^2\, q_{\alpha,k,\nu}^2}{L^2} = \frac{4 \times 15.95 \times 4.30^2}{10.0^2} = 11.8$$

so that an additional sample size of $12 - 6 = 6$ observations from each factor level can be recommended.

11.1.11   (a) $\bar{x}_{1.} = 5.633$

           $\bar{x}_{2.} = 5.567$

           $\bar{x}_{3.} = 4.778$

   (b) $\bar{x}_{..} = 5.326$

   (c) $SSTR = 4.076$

   (d) $\sum_{i=1}^{k} \sum_{j=1}^{n_i} x_{ij}^2 = 791.30$

   (e) $SST = 25.432$

   (f) $SSE = 21.356$

   (g)

Source	df	SS	MS	F	p-value
Treatments	2	4.076	2.038	2.29	0.123
Error	24	21.356	0.890		
Total	26	25.432			

   (h) $\mu_1 - \mu_2 \in \left(5.633 - 5.567 - \frac{\sqrt{0.890 \times 3.53}}{\sqrt{9}}, 5.633 - 5.567 + \frac{\sqrt{0.890 \times 3.53}}{\sqrt{9}}\right)$

         $= (-1.04, 1.18)$

        $\mu_1 - \mu_3 \in \left(5.633 - 4.778 - \frac{\sqrt{0.890 \times 3.53}}{\sqrt{9}}, 5.633 - 4.778 + \frac{\sqrt{0.890 \times 3.53}}{\sqrt{9}}\right)$

         $= (-0.25, 1.97)$

        $\mu_2 - \mu_3 \in \left(5.567 - 4.778 - \frac{\sqrt{0.890 \times 3.53}}{\sqrt{9}}, 5.567 - 4.778 + \frac{\sqrt{0.890 \times 3.53}}{\sqrt{9}}\right)$

         $= (-0.32, 1.90)$

   (j)   The total sample size required from each factor level can be estimated as

$$n \geq \frac{4\,s^2\,q_{\alpha,k,\nu}^2}{L^2} = \frac{4 \times 0.890 \times 3.53^2}{1.0^2} = 44.4$$

so that an additional sample size of $45 - 9 = 36$ observations from each factor level can be recommended.

Note: In the remainder of this section the confidence intervals for the pairwise differences of the factor level means are provided with an overall confidence level of 95%.

11.1.13

Source	df	SS	MS	F	p-value
Treatments	2	0.0085	0.0042	0.24	0.787
Error	87	1.5299	0.0176		
Total	89	1.5384			

$\mu_1 - \mu_2 \in (-0.08, 0.08)$

$\mu_1 - \mu_3 \in (-0.06, 0.10)$

$\mu_2 - \mu_3 \in (-0.06, 0.10)$

There is *not* sufficient evidence to conclude that there is a difference between the three production lines.

11.1.15

Source	df	SS	MS	F	p-value
Treatments	2	0.0278	0.0139	1.26	0.299
Error	30	0.3318	0.0111		
Total	32	0.3596			

$\mu_1 - \mu_2 \in (-0.15, 0.07)$

$\mu_1 - \mu_3 \in (-0.08, 0.14)$

$\mu_2 - \mu_3 \in (-0.04, 0.18)$

There is *not* sufficient evidence to conclude that the radiation readings are affected by the background radiation levels.

11.1.17

Source	df	SS	MS	F	p-value
Treatments	2	0.4836	0.2418	7.13	0.001
Error	93	3.1536	0.0339		
Total	95	3.6372			

$\mu_1 - \mu_2 \in (-0.01, 0.22)$

$\mu_1 - \mu_3 \in (0.07, 0.29)$

$\mu_2 - \mu_3 \in (-0.03, 0.18)$

There is sufficient evidence to conclude that the average particle diameter is larger at the low amount of stabilizer than at the high amount of stabilizer.

11.1.19 $\quad \bar{x}_{..} = \frac{(8 \times 42.91) + (11 \times 44.03) + (10 \times 43.72)}{8 + 11 + 10} = \frac{1264.81}{29} = 43.61$

$SSTr = (8 \times 42.91^2) + (11 \times 44.03^2) + (10 \times 43.72^2) - \frac{1264.81^2}{29} = 5.981$

$SSE = (7 \times 5.33^2) + (10 \times 4.01^2) + (9 \times 5.10^2) = 593.753$

Source	df	SS	MS	F	p-value
Treatments	2	5.981	2.990	0.131	0.878
Error	26	593.753	22.837		
Total	28	599.734			

There is not sufficient evidence to conclude that there is a difference between the catalysts in terms of the strength of the compound.

11.1.21 $\quad q_{0.05,5,43} = 4.04$

With a 95% confidence level the pairwise confidence intervals that contain zero are:

$\mu_1 - \mu_2$

$\mu_2 - \mu_5$

$\mu_3 - \mu_4$

It can be inferred that the largest mean is either μ_3 or μ_4

and that the smallest mean is either μ_2 or μ_5.

11.1.23 $\quad \bar{x}_{1.} = 40.80$

$\bar{x}_{2.} = 32.80$

$\bar{x}_{3.} = 25.60$

$\bar{x}_{4.} = 50.60$

$\bar{x}_{5.} = 41.80$

$\bar{x}_{6.} = 31.80$

Source	df	SS	MS	F	p-value
Physician	5	1983.8	396.8	15.32	0.000
Error	24	621.6	25.9		
Total	29	2605.4			

The p-value of 0.000 implies that there is sufficient evidence to conclude that the times taken by the physicians for the investigatory surgical procedures are different.

Since

$$\frac{s \times q_{0.05,6,24}}{\sqrt{5}} = \frac{\sqrt{25.9} \times 4.37}{\sqrt{5}} = 9.95$$

it follows that two physicians cannot be concluded to be different if their sample averages have a difference of less than 9.95.

The slowest physician is either physician 1, physician 4, or physician 5.

The quickest physician is either physician 2, physician 3, or physician 6.

11.1.25   (a) $\bar{x}_{1.} = 46.83$

$\bar{x}_{2.} = 47.66$

$\bar{x}_{3.} = 48.14$

$\bar{x}_{4.} = 48.82$

$\bar{x}_{..} = 47.82$

$SSTr = \sum_{i=1}^{4} n_i (\bar{x}_{i.} - \bar{x}_{..})^2 = 13.77$

Since the p-value is 0.01, the F-statistic in the analysis of variance table must be $F_{0.01,3,24} = 4.72$ so that the complete analysis of variance table is

Source	df	SS	MS	F	p-value
Treatments	3	13.77	4.59	4.72	0.01
Error	24	23.34	0.97		
Total	27	37.11			

(b) With $s = \sqrt{MSE} = 0.986$ and $q_{0.05,4,24} = 3.90$ the pairwise confidence intervals for the treatment means are:

$\mu_1 - \mu_2 \in (-2.11, 0.44)$

$\mu_1 - \mu_3 \in (-2.63, 0.01)$

$\mu_1 - \mu_4 \in (-3.36, -0.62)$

$\mu_2 - \mu_3 \in (-1.75, 0.79)$

$\mu_2 - \mu_4 \in (-2.48, 0.18)$

$\mu_3 - \mu_4 \in (-2.04, 0.70)$

There is sufficient evidence to establish that μ_4 is larger than μ_1.

11.1.27　B

11.2   Randomized Block Designs

11.2.1

Source	df	SS	MS	F	p-value
Treatments	3	10.15	3.38	3.02	0.047
Blocks	9	24.53	2.73	2.43	0.036
Error	27	30.24	1.12		
Total	39	64.92			

11.2.3

Source	df	SS	MS	F	p-value
Treatments	3	58.72	19.57	0.63	0.602
Blocks	9	2839.97	315.55	10.17	0.0000
Error	27	837.96	31.04		
Total	39	3736.64			

11.2.5   (a)

Source	df	SS	MS	F	p-value
Treatments	2	8.17	4.085	8.96	0.0031
Blocks	7	50.19	7.17	15.72	0.0000
Error	14	6.39	0.456		
Total	23	64.75			

(b) $\mu_1 - \mu_2 \in \left(5.93 - 4.62 - \frac{\sqrt{0.456} \times 3.70}{\sqrt{8}}, 5.93 - 4.62 + \frac{\sqrt{0.456} \times 3.70}{\sqrt{8}}\right)$

$= (0.43, 2.19)$

$\mu_1 - \mu_3 \in \left(5.93 - 4.78 - \frac{\sqrt{0.456} \times 3.70}{\sqrt{8}}, 5.93 - 4.78 + \frac{\sqrt{0.456} \times 3.70}{\sqrt{8}}\right)$

$= (0.27, 2.03)$

$\mu_2 - \mu_3 \in \left(4.62 - 4.78 - \frac{\sqrt{0.456} \times 3.70}{\sqrt{8}}, 4.62 - 4.78 + \frac{\sqrt{0.456} \times 3.70}{\sqrt{8}}\right)$

$= (-1.04, 0.72)$

11.2.7   (a) $\bar{x}_{1.} = 6.0617$

$\bar{x}_{2.} = 7.1967$

$\bar{x}_{3.} = 5.7767$

(b) $\bar{x}_{.1} = 7.4667$

$\bar{x}_{.2} = 5.2667$

$\bar{x}_{.3} = 5.1133$

$\bar{x}_{.4} = 7.3300$

$\bar{x}_{.5} = 6.2267$

$\bar{x}_{.6} = 6.6667$

(c) $\bar{x}_{..} = 6.345$

(d) $SSTr = 6.7717$

(e) $SSBl = 15.0769$

(f) $\sum_{i=1}^{3} \sum_{j=1}^{6} x_{ij}^2 = 752.1929$

(g) $SST = 27.5304$

(h) $SSE = 5.6818$

(i)

Source	df	SS	MS	F	p-value
Treatments	2	6.7717	3.3859	5.96	0.020
Blocks	5	15.0769	3.0154	5.31	0.012
Error	10	5.6818	0.5682		
Total	17	27.5304			

(j) $\mu_1 - \mu_2 \in \left(6.06 - 7.20 - \frac{\sqrt{0.5682} \times 3.88}{\sqrt{6}}, 6.06 - 7.20 + \frac{\sqrt{0.5682} \times 3.88}{\sqrt{6}}\right)$

$= (-2.33, 0.05)$

$\mu_1 - \mu_3 \in \left(6.06 - 5.78 - \frac{\sqrt{0.5682} \times 3.88}{\sqrt{6}}, 6.06 - 5.78 + \frac{\sqrt{0.5682} \times 3.88}{\sqrt{6}}\right)$

$= (-0.91, 1.47)$

$\mu_2 - \mu_3 \in \left(7.20 - 5.78 - \frac{\sqrt{0.5682} \times 3.88}{\sqrt{6}}, 7.20 - 5.78 + \frac{\sqrt{0.5682} \times 3.88}{\sqrt{6}}\right)$

$= (0.22, 2.61)$

(1) The total sample size required from each factor level (number of blocks) can be estimated as

$$n \geq \frac{4 \, s^2 \, q_{\alpha,k,\nu}^2}{L^2} = \frac{4 \times 0.5682 \times 3.88^2}{2.0^2} = 8.6$$

so that an additional $9 - 6 = 3$ blocks can be recommended.

Note: In the remainder of this section the confidence intervals for the pairwise differences of the factor level means are provided with an overall confidence level of 95%.

11.2.9

Source	df	SS	MS	F	p-value
Treatments	2	17.607	8.803	2.56	0.119
Blocks	6	96.598	16.100	4.68	0.011
Error	12	41.273	3.439		
Total	20	155.478			

$\mu_1 - \mu_2 \in (-1.11, 4.17)$

$\mu_1 - \mu_3 \in (-0.46, 4.83)$

$\mu_2 - \mu_3 \in (-1.99, 3.30)$

There is *not* sufficient evidence to conclude that the calciners are operating at different efficiencies.

11.2.11

Source	df	SS	MS	F	p-value
Treatments	3	3231.2	1,077.1	4.66	0.011
Blocks	8	29256.1	3,657.0	15.83	0.000
Error	24	5545.1	231.0		
Total	35	38032.3			

$\mu_1 - \mu_2 \in (-8.20, 31.32)$

$\mu_1 - \mu_3 \in (-16.53, 22.99)$

$\mu_1 - \mu_4 \in (-34.42, 5.10)$

$$\mu_2 - \mu_3 \in (-28.09, 11.43)$$

$$\mu_2 - \mu_4 \in (-45.98, -6.46)$$

$$\mu_3 - \mu_4 \in (-37.65, 1.87)$$

There is sufficient evidence to conclude that driver 4 is better than driver 2.

11.2.13

Source	df	SS	MS	F	p-value
Treatments	4	8.462×10^8	2.116×10^8	66.55	0.000
Blocks	11	19.889×10^8	1.808×10^8	56.88	0.000
Error	44	1.399×10^8	3.179×10^6		
Total	59	29.750×10^8			

$$\mu_1 - \mu_2 \in (4372, 8510)$$

$$\mu_1 - \mu_3 \in (4781, 8919)$$

$$\mu_1 - \mu_4 \in (5438, 9577)$$

$$\mu_1 - \mu_5 \in (-3378, 760)$$

$$\mu_2 - \mu_3 \in (-1660, 2478)$$

$$\mu_2 - \mu_4 \in (-1002, 3136)$$

$$\mu_2 - \mu_5 \in (-9819, -5681)$$

$$\mu_3 - \mu_4 \in (-1411, 2727)$$

$$\mu_3 - \mu_5 \in (-10228, -6090)$$

$$\mu_4 - \mu_5 \in (-10886, -6748)$$

There is sufficient evidence to conclude that either agent 1 or agent 5 is the best agent.

The worst agent is either agent 2, 3 or 4.

11.2.15   (a)

Source	df	SS	MS	F	p-value
Treatments	3	0.151	0.0503	5.36	0.008
Blocks	6	0.324	0.054	5.75	0.002
Error	18	0.169	0.00939		
Total	27	0.644			

(b) With $q_{0.05,4,18} = 4.00$ and

$$\frac{\sqrt{MSE} \times q_{0.05,4,18}}{\sqrt{b}} = \frac{\sqrt{0.00939} \times 4.00}{\sqrt{7}} = 0.146$$

the pairwise confidence intervals are:

$$\mu_2 - \mu_1 \in 0.630 - 0.810 \pm 0.146 = (-0.326, -0.034)$$

$$\mu_2 - \mu_3 \in 0.630 - 0.797 \pm 0.146 = (-0.313, -0.021)$$

$$\mu_2 - \mu_4 \in 0.630 - 0.789 \pm 0.146 = (-0.305, -0.013)$$

None of these confidence intervals contains zero so there is sufficient evidence to conclude that treatment 2 has a smaller mean value than each of the other treatments.

11.2.17   The new analysis of variance table is

Source	df	SS	MS	F	p-value
Treatments	same	a^2 SSTr	a^2 MSTr	same	same
Blocks	same	a^2 SSBl	a^2 MSBl	same	same
Error	same	a^2 SSE	a^2 MSE		
Total	same	a^2 SST			

11.2.19 $\bar{x}_{1.} = 23.18$

$\bar{x}_{2.} = 23.58$

$\bar{x}_{3.} = 23.54$

$\bar{x}_{4.} = 22.48$

Source	df	SS	MS	F	p-value
Locations	3	3.893	1.298	0.49	0.695
Time	4	472.647	118.162	44.69	0.000
Error	12	31.729	2.644		
Total	19	508.270			

The p-value of 0.695 implies that there is not sufficient evidence to conclude that the pollution levels are different at the four locations.

The confidence intervals for all of the pairwise comparisons contain zero, so the graphical representation has one line joining all four sample means.

Chapter 12

Simple Linear Regression and Correlation

12.1 The Simple Linear Regression Model

12.1.1   (a) $4.2 + (1.7 \times 10) = 21.2$

         (b) $3 \times 1.7 = 5.1$

         (c) $P(N(4.2 + (1.7 \times 5), 3.2^2) \geq 12) = 0.587$

         (d) $P(N(4.2 + (1.7 \times 8), 3.2^2) \leq 17) = 0.401$

         (e) $P(N(4.2 + (1.7 \times 6), 3.2^2) \geq N(4.2 + (1.7 \times 7), 3.2^2)) = 0.354$

12.1.3   (a) $y = 5 + (0.9 \times 20) = 23.0$

         (b)   The expected value of the porosity decreases by $5 \times 0.9 = 4.5$.

         (c) $P(N(5 + (0.9 \times 25), 1.4^2) \leq 30) = 0.963$

         (d) $P\left(17 \leq N\left(5 + (0.9 \times 15), \frac{1.4^2}{4}\right) \leq 20\right) = 0.968$

12.1.5 $P(N(675.30 - (5.87 \times 80), 7.32^2) \leq 220)$

$= P\left(N(0,1) \leq \frac{220 - 205.7}{7.32}\right)$

$= \Phi(1.954) = 0.975$

12.1.7   B

12.2 Fitting the Regression Line

12.2.3 $\hat{\beta}_0 = 39.5$

         $\hat{\beta}_1 = -2.04$

         $\hat{\sigma}^2 = 17.3$

         $39.5 + (-2.04 \times (-2.0)) = 43.6$

12.2.5   (a) $\hat{\beta}_0 = 36.19$

              $\hat{\beta}_1 = 0.2659$

       (b) $\hat{\sigma}^2 = 70.33$

       (c)   Yes, since $\hat{\beta}_1 > 0$.

       (d) $36.19 + (0.2659 \times 72) = 55.33$

12.2.7   (a) $\hat{\beta}_0 = -29.59$

              $\hat{\beta}_1 = 0.07794$

       (b) $-29.59 + (0.07794 \times 2,600) = 173.1$

       (c) $0.07794 \times 100 = 7.794$

       (d) $\hat{\sigma}^2 = 286$

12.2.9   (a) $\hat{\beta}_0 = 12.864$

              $\hat{\beta}_1 = 0.8051$

       (b) $12.864 + (0.8051 \times 69) = 68.42$

       (c) $0.8051 \times 5 = 4.03$

(d) $\hat{\sigma}^2 = 3.98$

12.2.11   C

12.3   Inferences on the Slope Parameter $\hat{\beta}_1$

12.3.1   (a) $(0.522 - (2.921 \times 0.142), 0.522 + (2.921 \times 0.142)) = (0.107, 0.937)$

(b)   The t-statistic is

$\frac{0.522}{0.142} = 3.68$

and the p-value is 0.002.

12.3.3   (a) $s.e.(\hat{\beta}_1) = 0.08532$

(b) $(1.003 - (2.145 \times 0.08532), 1.003 + (2.145 \times 0.08532)) = (0.820, 1.186)$

(c)   The t-statistic is

$\frac{1.003}{0.08532} = 11.76$

and the p-value is 0.000.

12.3.5   (a) $s.e.(\hat{\beta}_1) = 0.1282$

(b) $(-\infty, -0.3377 + (1.734 \times 0.1282)) = (-\infty, -0.115)$

(c)   The t-statistic is

$\frac{-0.3377}{0.1282} = -2.63$

and the (two-sided) p-value is 0.017.

12.3.7   (a) $s.e.(\hat{\beta}_1) = 0.2829$

(b) $(1.619 - (2.042 \times 0.2829), 1.619 + (2.042 \times 0.2829)) = (1.041, 2.197)$

(c)   If $\beta_1 = 1$ then the actual times are equal to the estimated times except for a constant difference of β_0.

The t-statistic is

$$\frac{1.619 - 1.000}{0.2829} = 2.19$$

and the p-value is 0.036.

12.3.9   For the hypotheses

$H_0 : \beta_1 = 0$ versus $H_A : \beta_1 \neq 0$

the t-statistic is

$t = \frac{54.87}{21.20} = 2.588$

so that the p-value is $2 \times P(t_{18} \geq 2.588) = 0.019$.

12.3.11   A

12.4   Inferences on the Regression Line

12.4.3 $(21.9, 23.2)$

12.4.5 $(33.65, 41.02)$

12.4.7 $(-\infty, 10.63)$

12.4.9 $\sum_{i=1}^{8} x_i = 122.6$

$\sum_{i=1}^{8} x_i^2 = 1939.24$

$\bar{x} = \frac{122.6}{8} = 15.325$

$S_{XX} = 1939.24 - \frac{122.6^2}{8} = 60.395$

With $t_{0.025,6} = 2.447$ the confidence interval is

$\beta_0 + (\beta_1 \times 15) \in 75.32 + (0.0674 \times 15) \pm 2.447 \times 0.0543 \times \sqrt{\frac{1}{8} + \frac{(15-15.325)^2}{60.395}}$

which is $(76.284, 76.378)$.

12.5 Prediction Intervals for Future Response Values

12.5.1 $(1386, 1406)$

12.5.3 $(5302, 9207)$

12.5.5 $(165.7, 274.0)$

12.5.7 $(63.48, 74.96)$

12.5.9 $n = 7$

$\sum_{i=1}^{7} x_i = 142.8$

$\sum_{i=1}^{7} y_i = 361.5$

$\sum_{i=1}^{7} x_i^2 = 2942.32$

$\sum_{i=1}^{7} y_i^2 = 18771.5$

$\sum_{i=1}^{7} x_i y_i = 7428.66$

$\bar{x} = \frac{142.8}{7} = 20.400$

$\bar{y} = \frac{361.5}{7} = 51.643$

$S_{XX} = 2942.32 - \frac{142.8^2}{7} = 29.200$

$S_{YY} = 18771.5 - \frac{361.5^2}{7} = 102.607$

$S_{XY} = 7428.66 - \frac{142.8 \times 361.5}{7} = 54.060$

Using these values

$\hat{\beta}_1 = \frac{54.060}{29.200} = 1.851$

$\hat{\beta}_0 = 51.643 - (1.851 \times 20.400) = 13.875$

and

$$SSE = 18771.5 - (13.875 \times 361.5) - (1.851 \times 7428.66) = 2.472$$

so that

$$\hat{\sigma}^2 = \frac{2.472}{7-2} = 0.494.$$

With $t_{0.005,5} = 4.032$ the prediction interval is

$$13.875 + (1.851 \times 20) \pm 4.032 \times \sqrt{0.494} \times \sqrt{\frac{8}{7} + \frac{(20-20.400)^2}{29.200}}$$

which is

$$50.902 \pm 3.039 = (47.864, 53.941).$$

12.6 The Analysis of Variance Table

12.6.1

Source	df	SS	MS	F	p-value
Regression	1	40.53	40.53	2.32	0.137
Error	33	576.51	17.47		
Total	34	617.04			

$R^2 = \frac{40.53}{617.04} = 0.066$

12.6.3

Source	df	SS	MS	F	p-value
Regression	1	870.43	870.43	889.92	0.000
Error	8	7.82	0.9781		
Total	9	878.26			

$R^2 = \frac{870.43}{878.26} = 0.991$

12.6.5

Source	df	SS	MS	F	p-value
Regression	1	10.71×10^7	10.71×10^7	138.29	0.000
Error	14	1.08×10^7	774,211		
Total	15	11.79×10^7			

$R^2 = \frac{10.71 \times 10^7}{11.79 \times 10^7} = 0.908$

12.6.7

Source	df	SS	MS	F	p-value
Regression	1	397.58	397.58	6.94	0.017
Error	18	1031.37	57.30		
Total	19	1428.95			

$R^2 = \frac{397.58}{1428.95} = 0.278$

The R^2 value implies that about 28% of the variability in VO2-max can be accounted for by changes in age.

12.6.9

Source	df	SS	MS	F	p-value
Regression	1	411.26	411.26	32.75	0.000
Error	30	376.74	12.56		
Total	31	788.00			

$$R^2 = \frac{411.26}{788.00} = 0.522$$

The p-value is not very meaningful because it tests the null hypothesis that the actual times are unrelated to the estimated times.

12.6.11   A

12.7 Residual Analysis

12.7.1 There is no suggestion that the fitted regression model is not appropriate.

12.7.3 There is a possible suggestion of a slight reduction in the variability of the VO2-max values as age increases.

12.7.5 The variability of the actual times increases as the estimated time increases.

12.7.7 D

12.8   Variable Transformations

12.8.1   The model

$$y = \gamma_0 \, e^{\gamma_1 x}$$

is appropriate.

A linear regression can be performed with $\ln(y)$ as the dependent variable and with x as the input variable.

$\hat{\gamma}_0 = 9.12$

$\hat{\gamma}_1 = 0.28$

$\hat{\gamma}_0 \, e^{\hat{\gamma}_1 \times 2.0} = 16.0$

12.8.3 $\hat{\gamma}_0 = 8.81$

$\hat{\gamma}_1 = 0.523$

$\gamma_0 \in (6.84, 11.35)$

$\gamma_1 \in (0.473, 0.573)$

12.8.5 $\hat{\gamma}_0 = e^{\hat{\beta}_0} = e^{2.628} = 13.85$

$\hat{\gamma}_1 = \hat{\beta}_1 = 0.341$

With $t_{0.025,23} = 2.069$ the confidence interval for γ_1 (and β_1) is
$0.341 \pm (2.069 \times 0.025) = (0.289, 0.393)$.

12.8.7 $\hat{\gamma}_0 = 12.775$

$\hat{\gamma}_1 = -0.5279$

When the crack length is 2.1 the expected breaking load is

$12.775 \times e^{-0.5279 \times 2.1} = 4.22$.

12.9 Correlation Analysis

12.9.3 The sample correlation coefficient is $r = 0.95$.

12.9.5 The sample correlation coefficient is $r = -0.53$.

12.9.7 The sample correlation coefficient is $r = 0.72$.

12.9.9 The sample correlation coefficient is $r = 0.431$.

12.9.11 The variables A and B may both be related to a third surrogate variable C. It is possible that the variables A and C have a causal relationship, and that the variables B and C have a causal relationship, without there being a causal relationship between the variables A and B.

12.9.13 A

Chapter 13

Multiple Linear Regression

13.1 Introduction to Multiple Linear Regression

13.1.1   (a) $R^2 = 0.89$

        (b)

Source	df	SS	MS	F	p-value
Regression	3	96.5	32.17	67.4	0.000
Error	26	12.4	0.477		
Total	29	108.9			

        (c) $\hat{\sigma}^2 = 0.477$

        (d)   The p-value is 0.000.

        (e) $(16.5 - (2.056 \times 2.6), 16.5 + (2.056 \times 2.6)) = (11.2, 21.8)$

13.1.3   (a) $(132.4 - (2.365 \times 27.6), 132.4 + (2.365 \times 27.6)) = (67.1, 197.7)$

        (b)   The test statistic is $t = 4.80$ and the p-value is 0.002.

13.1.5   The test statistic for $H_0 : \beta_1 = 0$ is $t = 11.30$ and the p-value is 0.000.

        The test statistic for $H_0 : \beta_2 = 0$ is $t = 5.83$ and the p-value is 0.000.

        The test statistic for $H_0 : \beta_3 = 0$ is $t = 1.15$ and the p-value is 0.257.

Variable x_3 should be removed from the model.

13.1.7   The test statistic is $F = 5.29$ and the p-value is 0.013.

13.1.9   (a) $\hat{y} = 104.9 + (12.76 \times 10) + (409.6 \times 0.3) = 355.38$

(b) $(355.38 - (2.110 \times 17.6), 355.38 + (2.110 \times 17.6)) = (318.24, 392.52)$

13.1.11 $MSE = 4.33^2 = 18.749$

$SST = 694.09 - \frac{(-5.68)^2}{20} = 692.477$

Source	df	SS	MS	F	p-value
Regression	3	392.495	130.832	6.978	0.003
Error	16	299.982	18.749		
Total	19	692.477			

The p-value in the analysis of variance table is for the null hypothesis $H_0 : \beta_1 = \beta_2 = \beta_3 = 0$.

The proportion of the variability of the y variable that is explained by the model is

$R^2 = \frac{392.495}{692.477} = 56.7\%$.

13.1.13   B

13.1.15   A

13.1.17   B

13.2 Examples of Multiple Linear Regression

13.2.1   (b) The variable competitor's price has a p-value of 0.216 and is not needed in the model.

The sample correlation coefficient between the competitor's price and the sales is $r = -0.91$.

The sample correlation coefficient between the competitor's price and the company's price is $r = 0.88$.

(c) The sample correlation coefficient between the company's price and the sales is $r = -0.96$.

Using the model

sales $= 107.4 - (3.67 \times$ company's price$)$

the predicted sales are $107.4 - (3.67 \times 10.0) = 70.7$.

13.2.3   (a) $\hat{\beta}_0 = -3,238.6$

$\hat{\beta}_1 = 0.9615$

$\hat{\beta}_2 = 0.732$

$\hat{\beta}_3 = 2.889$

$\hat{\beta}_4 = 389.9$

(b) The variable geology has a p-value of 0.737 and is not needed in the model.

The sample correlation coefficient between the cost and geology is $r = 0.89$.

The sample correlation coefficient between the depth and geology is $r = 0.92$.

The variable geology is not needed in the model because it is highly correlated with the variable depth which is in the model.

(c) The variable rig-index can also be removed from the model.

A final model

$$\text{cost} = -3011 + (1.04 \times \text{depth}) + (2.67 \times \text{downtime})$$

can be recommended.

13.2.5 Two indicator variables x_1 and x_2 are needed.
One way is to have $(x_1, x_2) = (0, 0)$ at level 1,
$(x_1, x_2) = (0, 1)$ at level 2,
and $(x_1, x_2) = (1, 0)$ at level 3.

13.3   Matrix Algebra Formulation

13.3.1   (a)

$$\mathbf{Y} = \begin{pmatrix} 2 \\ -2 \\ 4 \\ -2 \\ 2 \\ -4 \\ 1 \\ 3 \\ 1 \\ -5 \end{pmatrix}$$

(b)

$$\mathbf{X} = \begin{pmatrix} 1 & 0 & 1 \\ 1 & 0 & -1 \\ 1 & 1 & 4 \\ 1 & 1 & -4 \\ 1 & -1 & 2 \\ 1 & -1 & -2 \\ 1 & 2 & 0 \\ 1 & 2 & 0 \\ 1 & -2 & 3 \\ 1 & -2 & -3 \end{pmatrix}$$

(c)

$$\mathbf{X'X} = \begin{pmatrix} 10 & 0 & 0 \\ 0 & 20 & 0 \\ 0 & 0 & 60 \end{pmatrix}$$

(d)

$$(\mathbf{X'X})^{-1} = \begin{pmatrix} 0.1000 & 0 & 0 \\ 0 & 0.0500 & 0 \\ 0 & 0 & 0.0167 \end{pmatrix}$$

(e)

$$\mathbf{X'Y} = \begin{pmatrix} 0 \\ 20 \\ 58 \end{pmatrix}$$

(g)

$$\hat{\mathbf{Y}} = \begin{pmatrix} 0.967 \\ -0.967 \\ 4.867 \\ -2.867 \\ 0.933 \\ -2.933 \\ 2.000 \\ 2.000 \\ 0.900 \\ -4.900 \end{pmatrix}$$

(h)

$$\mathbf{e} = \begin{pmatrix} 1.033 \\ -1.033 \\ -0.867 \\ 0.867 \\ 1.067 \\ -1.067 \\ -1.000 \\ 1.000 \\ 0.100 \\ -0.100 \end{pmatrix}$$

(i) $SSE = 7.933$

(k) $s.e.(\hat{\beta}_1) = 0.238$

$s.e.(\hat{\beta}_2) = 0.137$

Both input variables should be kept in the model.

(l) The fitted value is

$$0 + (1 \times 1) + \left(\tfrac{29}{30} \times 2 \right) = 2.933.$$

The standard error is 0.496.

The confidence interval is $(1.76, 4.11)$.

(m) The prediction interval is $(0.16, 5.71)$.

13.3.3

$$\mathbf{Y} = \begin{pmatrix} 10 \\ 0 \\ -5 \\ 2 \\ 3 \\ -6 \end{pmatrix}$$

$$\mathbf{X} = \begin{pmatrix} 1 & -3 & 1 & 3 \\ 1 & -2 & 1 & 0 \\ 1 & -1 & 1 & -5 \\ 1 & 1 & -6 & 1 \\ 1 & 2 & -3 & 0 \\ 1 & 3 & 6 & 1 \end{pmatrix}$$

$$\mathbf{X}'\mathbf{X} = \begin{pmatrix} 6 & 0 & 0 & 0 \\ 0 & 28 & 0 & 0 \\ 0 & 0 & 84 & -2 \\ 0 & 0 & -2 & 36 \end{pmatrix}$$

$$(\mathbf{X}'\mathbf{X})^{-1} = \begin{pmatrix} 0.16667 & 0 & 0 & 0 \\ 0 & 0.03571 & 0 & 0 \\ 0 & 0 & 0.01192 & 0.00066 \\ 0 & 0 & 0.00066 & 0.02781 \end{pmatrix}$$

$$\mathbf{X}'\mathbf{Y} = \begin{pmatrix} 4 \\ -35 \\ -52 \\ 51 \end{pmatrix}$$

$$\hat{\beta} = \begin{pmatrix} 0.6667 \\ -1.2500 \\ -0.5861 \\ 1.3841 \end{pmatrix}$$

13.4 Evaluating Model Accuracy

13.4.1   (a)   There is a slight suggestion of a greater variability in the yields at higher temperatures.

         (b)   There are no unusually large standardized residuals.

         (c)   The points $(90, 85)$ and $(200, 702)$ have leverage values $h_{ii} = 0.547$.

13.4.3   (a)   The residual plots do not indicate any substantial problems.

         (b)   If it were beneficial to add the variable weight to the model, then there would be some pattern in this residual plot.

         (d)   The observation with VO2-max $= 23$ has a standardized residual of -2.15.

13.4.5   B

13.4.7   B

Chapter 14

Multifactor Experimental Design

14.1 Experiments with Two Factors

14.1.1

Source	df	SS	MS	F	p-value
Fuel	1	96.33	96.33	3.97	0.081
Car	1	75.00	75.00	3.09	0.117
Fuel*Car	1	341.33	341.33	14.08	0.006
Error	8	194.00	24.25		
Total	11	706.66			

14.1.3 (a)

Source	df	SS	MS	F	p-value
Tip	2	0.1242	0.0621	1.86	0.175
Material	2	14.1975	7.0988	212.31	0.000
Tip*Material	4	0.0478	0.0120	0.36	0.837
Error	27	0.9028	0.0334		
Total	35	15.2723			

(c) Apart from one large negative residual there appears to be less variability in the measurements from the third tip.

166

14.1.5

Source	df	SS	MS	F	p-value
Glass	2	7.568	3.784	0.70	0.514
Acidity	1	12.667	12.667	2.36	0.151
Glass*Acidity	2	93.654	46.827	8.71	0.005
Error	12	64.540	5.378		
Total	17	178.429			

14.1.7

Source	df	SS	MS	F	p-value
Design	2	3.896×10^3	1.948×10^3	0.46	0.685
Material	1	0.120×10^3	0.120×10^3	0.03	0.882
Error	2	8.470×10^3	4.235×10^3		
Total	5	12.487×10^3			

14.2   Experiments with Three or More Factors

14.2.1   (d)

Source	df	SS	MS	F	p-value
Drink	2	90.65	45.32	5.39	0.007
Gender	1	6.45	6.45	0.77	0.384
Age	2	23.44	11.72	1.39	0.255
Drink*Gender	2	17.82	8.91	1.06	0.352
Drink*Age	4	24.09	6.02	0.72	0.583
Gender*Age	2	24.64	12.32	1.47	0.238
Drink*Gender*Age	4	27.87	6.97	0.83	0.511
Error	72	605.40	8.41		
Total	89	820.36			

14.2.3   (a)

Source	df	SS	MS	F	p-value
Add-A	2	324.11	162.06	8.29	0.003
Add-B	2	5.18	2.59	0.13	0.877
Conditions	1	199.28	199.28	10.19	0.005
Add-A*Add-B	4	87.36	21.84	1.12	0.379
Add-A*Conditions	2	31.33	15.67	0.80	0.464
Add-B*Conditions	2	2.87	1.44	0.07	0.930
Add-A*Add-B*Conditions	4	21.03	5.26	0.27	0.894
Error	18	352.05	19.56		
Total	35	1023.21			

The amount of additive B does not effect the expected value of the gas mileage although the variability of the gas mileage increases as more of additive B is used.

14.2.5 (d)

Source	df	SS	MS	F	p-value
Machine	1	387.1	387.1	3.15	0.095
Temp	1	29.5	29.5	0.24	0.631
Position	1	1271.3	1271.3	10.35	0.005
Angle	1	6865.0	6685.0	55.91	0.000
Machine*Temp	1	43.0	43.0	0.35	0.562
Machine*Position	1	54.9	54.9	0.45	0.513
Machine*Angle	1	1013.6	1013.6	8.25	0.011
Temp*Position	1	67.6	67.6	0.55	0.469
Temp*Angle	1	8.3	8.3	0.07	0.798
Position*Angle	1	61.3	61.3	0.50	0.490
Machine*Temp*Position	1	21.0	21.0	0.17	0.685
Machine*Temp*Angle	1	31.4	31.4	0.26	0.620
Machine*Position*Angle	1	13.7	13.7	0.11	0.743
Temp*Position*Angle	1	17.6	17.6	0.14	0.710
Machine*Temp*Position*Angle	1	87.5	87.5	0.71	0.411
Error	16	1964.7	122.8		
Total	31	11937.3			

14.2.7 A redundant

B redundant

C redundant

D redundant

A*B not significant

A*C redundant

A*D redundant

B*C significant

B*D significant

C*D redundant

A*B*C not significant

A*B*D not significant

A*C*D significant

B*C*D not significant

A*B*C*D not significant

Chapter 15

Nonparametric Statistical Analysis

15.1 The Analysis of a Single Population

15.1.1   (c) It is not plausible.

       (d) It is not plausible.

       (e) $S(65) = 84$
          The p-value is 0.064.

       (f) The p-value is 0.001.

       (g) The confidence interval from the sign test is $(65.0, 69.0)$.
          The confidence interval from the signed rank test is $(66.0, 69.5)$.

15.1.3   The p-values for the hypotheses

       $H_0 : \mu = 0.2$ versus $H_A : \mu \neq 0.2$

       are 0.004 for the sign test,

       0.000 for the signed rank test,

       and 0.000 for the t-test.

170

Confidence intervals for μ with a confidence level of at least 95% are

$(0.207, 0.244)$ for the sign test,

$(0.214, 0.244)$ for the signed rank test,

and $(0.216, 0.248)$ for the t-test.

There is sufficient evidence to conclude that the median paint thickness is larger than 0.2 mm.

15.1.5   (a) $S(18.0) = 14$

(b) The exact p-value is
$$2 \times P(B(20, 0.5) \geq 14) = 0.115.$$

(c) $2 \times \Phi(-1.57) = 0.116$

(d) $S_+(18.0) = 37$

(e) $2 \times \Phi(-2.52) = 0.012$

15.1.7   It is reasonable to assume that the differences of the data have a symmetric distribution in which case the signed rank test can be used.

The p-values for the hypotheses

$H_0 : \mu_A - \mu_B = 0$ versus $H_A : \mu_A - \mu_B \neq 0$

are 0.296 for the sign test and 0.300 for the signed rank test.

Confidence intervals for $\mu_A - \mu_B$ with a confidence level of at least 95% are

$(-1.0, 16.0)$ for the sign test and

$(-6.0, 17.0)$ for the signed rank test.

There is not sufficient evidence to conclude that there is a difference between the two assembly methods.

15.1.9   The p-values for the hypotheses

$H_0 : \mu_A - \mu_B = 0$ versus $H_A : \mu_A - \mu_B \neq 0$

are 0.003 for the sign test and 0.002 for the signed rank test.

Confidence intervals for $\mu_A - \mu_B$ with a confidence level of at least 95% are

$(-13.0, -1.0)$ for the sign test and

$(-12.0, -3.5)$ for the signed rank test.

The signed rank test shows that the new teaching method is better by at least 3.5 points on average.

15.1.11   The p-values for the hypotheses

$H_0 : \mu_A - \mu_B = 0$ versus $H_A : \mu_A - \mu_B \neq 0$

are 0.541 for the sign test and 0.721 for the signed rank test.

Confidence intervals for $\mu_A - \mu_B$ with a confidence level of at least 95% are

$(-13.6, 7.3)$ for the sign test and

$(-6.6, 6.3)$ for the signed rank test.

There is not sufficient evidence to conclude that there is a difference between the two ball types.

15.2 Comparing Two Populations

15.2.1   (c) The Kolmogorov-Smirnov statistic is $M = 0.2006$, which is larger than

$$d_{0.01} \sqrt{\tfrac{1}{200} + \tfrac{1}{180}} = 0.167.$$

There is sufficient evidence to conclude that the two distribution functions are different.

15.2.3   The Kolmogorov-Smirnov statistic is $M = 0.40$, which is larger than

$$d_{0.01} \sqrt{\tfrac{1}{50} + \tfrac{1}{50}} = 0.326.$$

There is sufficient evidence to conclude that the two distribution functions are different.

15.2.5   (b) $S_A = 245$

  (c) $U_A = 245 - \frac{14 \times (14+1)}{2} = 140$

  (d) Since

$$U_A = 140 > \frac{mn}{2} = \frac{14 \times 12}{2} = 84$$

the value of U_A is consistent with the observations from population A being larger than the observations from population B.

  (e) The p-value is 0.004.

There is sufficient evidence to conclude that the observations from population A tend to be larger than the observations from population B.

15.2.7   (c) The Kolmogorov-Smirnov statistic is $M = 0.218$, which is approximately equal to

$$d_{0.05} \sqrt{\tfrac{1}{75} + \tfrac{1}{82}} = 0.217.$$

There is some evidence that the two distribution functions are different, although the evidence is not overwhelming.

(d) $S_A = 6555.5$

$U_A = 6555.5 - \frac{75 \times (75+1)}{2} = 3705.5$

Since

$U_A = 3705.5 > \frac{mn}{2} = \frac{75 \times 82}{2} = 3075.0$

the value of U_A is consistent with the observations from production line A being larger than the observations from production line B.

The two-sided p-value is 0.027.

A 95% confidence interval for the difference in the population medians is
$(0.003, 0.052)$.

The rank sum test is based on the assumption that the two distribution functions are identical except for a location difference, and the plots of the empirical cumulative distribution functions in (a) suggest that this assumption is not unreasonable.

15.3 Comparing Three or More Populations

15.3.1 (b) $\bar{r}_{1.} = 16.6$

$\bar{r}_{2.} = 15.5$

$\bar{r}_{3.} = 9.9$

(c) $H = 3.60$

(d) The p-value is $P(\chi_2^2 > 3.60) = 0.165$.

15.3.3 (a) $\bar{r}_{1.} = 17.0$

$\bar{r}_{2.} = 19.8$

$\bar{r}_{3.} = 14.2$

$H = 1.84$

The p-value is $P(\chi_2^2 > 1.84) = 0.399$.

There is not sufficient evidence to conclude that the radiation readings are affected by the background radiation level.

(b) See Problem 11.1.15.

15.3.5 $\bar{r}_{1.} = 55.1$

$\bar{r}_{2.} = 55.7$

$\bar{r}_{3.} = 25.7$

$H = 25.86$

The p-value is $P(\chi_2^2 > 25.86) = 0.000$.

There is sufficient evidence to conclude that the computer assembly times are affected by the different assembly methods.

15.3.7   (a) $\bar{r}_{1.} = 2.250$

$\bar{r}_{2.} = 1.625$

$\bar{r}_{3.} = 3.500$

$\bar{r}_{4.} = 2.625$

(b) $S = 8.85$

(c) The p-value is $P(\chi^2_3 > 8.85) = 0.032$.

15.3.9 $\bar{r}_{1.} = 1.125$

$\bar{r}_{2.} = 2.875$

$\bar{r}_{3.} = 2.000$

$S = 12.25$

The p-value is $P(\chi^2_2 > 12.25) = 0.002$.

There is sufficient evidence to conclude that there is a difference between the radar systems.

15.3.11 $\bar{r}_{1.} = 4.42$

$\bar{r}_{2.} = 2.50$

$\bar{r}_{3.} = 1.79$

$\bar{r}_{4.} = 1.71$

$\bar{r}_{5.} = 4.58$

$S = 37.88$

The p-value is $P(\chi^2_4 > 37.88) = 0.000$.

There is sufficient evidence to conclude that there is a difference in the performances of the agents.

Chapter 16

Quality Control Methods

16.2 Statistical Process Control

16.2.1   (a) The center line is 10.0 and the control limits are 9.7 and 10.3.

      (b) The process is declared to be out of control at $\bar{x} = 9.5$ but not at $\bar{x} = 10.25$.

      (c) $P\left(9.7 \leq N\left(10.15, \frac{0.2^2}{4}\right) \leq 10.3\right) = 0.9332$

         The probability that an observation lies outside the control limits is therefore $1 - 0.9332 = 0.0668$.

         The average run length for detecting the change is $\frac{1}{0.0668} = 15.0$.

16.2.3   (a) $P(\mu - 2\sigma \leq N(\mu, \sigma^2) \leq \mu + 2\sigma) = 0.9544$

         The probability that an observation lies outside the control limits is therefore $1 - 0.9544 = 0.0456$.

      (b) $P(\mu - 2\sigma \leq N(\mu + \sigma, \sigma^2) \leq \mu + 2\sigma) = 0.8400$

         The probability that an observation lies outside the control limits is therefore $1 - 0.8400 = 0.1600$.

         The average run length for detecting the change is $\frac{1}{0.1600} = 6.25$.

177

16.2.5   The probability that a point is above the center line and within the upper
control limit is

$$P(\mu \leq N(\mu, \sigma^2) \leq \mu + 3\sigma) = 0.4987.$$

The probability that all eight points lie above the center line and within
the upper control limit is therefore $0.4987^8 = 0.0038$.

Similarly, the probability that all eight points lie below the center line
and within the lower control limit is $0.4987^8 = 0.0038$.

Consequently, the probability that all eight points lie on the same side of
the center line and within the control limits is $2 \times 0.0038 = 0.0076$.

Since this probability is very small, if all eight points lie on the same side
of the centerline this suggests that the process has moved out of control,
even though the points may all lie within the control limits.

16.3 Variable Control Charts

16.3.1   (a) The $\bar{X}$-chart has a center line at 91.33 and control limits at 87.42 and 95.24.

              The R-chart has a center line at 5.365 and control limits at 0 and 12.24.

   (b) No

   (c) $\bar{x} = 92.6$

          $r = 13.1$

          The process can be declared to be out of control due to an increase in the variability.

   (d) $\bar{x} = 84.6$

          $r = 13.5$

          The process can be declared to be out of control due to an increase in the variability and a decrease in the mean value.

   (e) $\bar{x} = 91.8$

          $r = 5.7$

          There is no evidence that the process is out of control.

   (f) $\bar{x} = 95.8$

          $r = 5.4$

          The process can be declared to be out of control due to an increase in the mean value.

16.3.3   (a) The $\bar{X}$-chart has a center line at 2.993 and control limits at 2.801 and 3.186.

              The R-chart has a center line at 0.2642 and control limits at 0 and 0.6029.

(b) $\bar{x} = 2.97$

$r = 0.24$

There is no evidence that the process is out of control.

16.4 Attribute Control Charts

16.4.1   The p-chart has a center line at 0.0500 and control limits at 0.0000 and 0.1154.

(a) No

(b) In order for

$$\tfrac{x}{100} \geq 0.1154$$

it is necessary that $x \geq 12$.

16.4.3   The c-chart has a center line at 12.42 and control limits at 1.85 and 22.99.

(a) No

(b) At least 23.

16.5   Acceptance Sampling

16.5.1   (a) With $p_0 = 0.06$ there would be 3 defective items in the batch of $N = 50$ items.

The producer's risk is 0.0005.

(b) With $p_1 = 0.20$ there would be 10 defective items in the batch of $N = 50$ items.

The consumer's risk is 0.952.

Using a binomial approximation these probabilities are estimated to be 0.002 and 0.942.

16.5.3   (a) The producer's risk is 0.000.

(b) The consumer's risk is 0.300.

16.5.5   The smallest value of c for which

$$P(B(30, 0.10) > c) \leq 0.05$$

is $c = 6$.

Chapter 17

Reliability Analysis and Life Testing

17.1 System Reliability

17.1.1 $r = 0.9985$

17.1.3   (a)   If r_1 is the individual reliability then in order for

$$r_1^4 \geq 0.95$$

it is necessary that $r_1 \geq 0.9873$.

(b)   If r_1 is the individual reliability then in order for

$$1 - (1 - r_1)^4 \geq 0.95$$

it is necessary that $r_1 \geq 0.5271$.

(c)   Suppose that n components with individual reliabilities r_1 are used, then an overall reliability of r is achieved as long as

$$r_1 \geq r^{1/n}$$

when the components are placed in series, and as long as

$$r_1 \geq 1 - (1 - r)^{1/n}$$

when the components are placed in parallel.

17.1.5 $r = 0.820$

17.2   Modeling Failure Rates

17.2.1   The parameter is $\lambda = \frac{1}{225}$.

(a) $P(T \geq 250) = e^{-250/225} = 0.329$

(b) $P(T \leq 150) = 1 - e^{-150/225} = 0.487$

(c) $P(T \geq 100) = e^{-100/225} = 0.641$

If three components are placed in series then the system reliability is $0.641^3 = 0.264$.

17.2.3 $\dfrac{1}{\frac{1}{125} + \frac{1}{60} + \frac{1}{150} + \frac{1}{100}} = 24.2$ minutes

17.2.5   (a) $P(T \geq 40) = 1 - \Phi\left(\frac{\ln(40) - 2.5}{1.5}\right) = 0.214$

(b) $P(T \leq 10) = \Phi\left(\frac{\ln(10) - 2.5}{1.5}\right) = 0.448$

(c) $e^{2.5 + 1.5^2/2} = 37.5$

(d) Solving

$$\Phi\left(\frac{\ln(t) - 2.5}{1.5}\right) = 0.5$$

gives $t = e^{2.5} = 12.2$.

17.2.7   (a) $P(T \geq 5) = e^{-(0.25 \times 5)^{3.0}} = 0.142$

(b) $P(T \leq 3) = 1 - e^{-(0.25 \times 3)^{3.0}} = 0.344$

(c) Solving

$$1 - e^{-(0.25 \times t)^{3.0}} = 0.5$$

gives $t = 3.54$.

(d) The hazard rate is

$$h(t) = 3.0 \times 0.25^{3.0} \times t^{3.0-1} = 0.0469 \times t^2.$$

(e) $\frac{h(5)}{h(3)} = 2.78$

17.3   Life Testing

17.3.1   (a)   With $\chi^2_{60,0.005} = 91.952$ and $\chi^2_{60,0.995} = 35.534$ the confidence interval is

$$\left(\frac{2 \times 30 \times 132.4}{91.952}, \frac{2 \times 30 \times 132.4}{35.534} \right) = (86.4, 223.6).$$

(b)   The value 150 is within the confidence interval, so the claim is plausible.

17.3.3   (a)   With $\chi^2_{60,0.005} = 91.952$ and $\chi^2_{60,0.995} = 35.534$, and with $\bar{t} = 176.5/30 = 5.883$,

the confidence interval is

$$\left(\frac{2 \times 30 \times 5.883}{91.952}, \frac{2 \times 30 \times 5.883}{35.534} \right) = (3.84, 9.93).$$

(b)   The value 10 is not included within the confidence interval, and so it is not plausible that the mean time to failure is 10 hours.

17.3.5　(a)

$$0 < t \le 67 \quad \Rightarrow \quad \hat{r}(t) = 1$$
$$67 < t \le 72 \quad \Rightarrow \quad \hat{r}(t) = 1 \times (27 - 1)/27 = 0.963$$
$$72 < t \le 79 \quad \Rightarrow \quad \hat{r}(t) = 0.963 \times (26 - 2)/26 = 0.889$$
$$79 < t \le 81 \quad \Rightarrow \quad \hat{r}(t) = 0.889 \times (24 - 1)/24 = 0.852$$
$$81 < t \le 82 \quad \Rightarrow \quad \hat{r}(t) = 0.852 \times (22 - 1)/22 = 0.813$$
$$82 < t \le 89 \quad \Rightarrow \quad \hat{r}(t) = 0.813 \times (21 - 1)/21 = 0.774$$
$$89 < t \le 93 \quad \Rightarrow \quad \hat{r}(t) = 0.774 \times (18 - 1)/18 = 0.731$$
$$93 < t \le 95 \quad \Rightarrow \quad \hat{r}(t) = 0.731 \times (17 - 1)/17 = 0.688$$
$$95 < t \le 101 \quad \Rightarrow \quad \hat{r}(t) = 0.688 \times (16 - 1)/16 = 0.645$$
$$101 < t \le 104 \quad \Rightarrow \quad \hat{r}(t) = 0.645 \times (15 - 1)/15 = 0.602$$
$$104 < t \le 105 \quad \Rightarrow \quad \hat{r}(t) = 0.602 \times (14 - 1)/14 = 0.559$$
$$105 < t \le 109 \quad \Rightarrow \quad \hat{r}(t) = 0.559 \times (13 - 1)/13 = 0.516$$
$$109 < t \le 114 \quad \Rightarrow \quad \hat{r}(t) = 0.516 \times (11 - 1)/11 = 0.469$$
$$114 < t \le 122 \quad \Rightarrow \quad \hat{r}(t) = 0.469 \times (9 - 2)/9 = 0.365$$
$$122 < t \le 126 \quad \Rightarrow \quad \hat{r}(t) = 0.365 \times (7 - 1)/7 = 0.313$$
$$126 < t \le 135 \quad \Rightarrow \quad \hat{r}(t) = 0.313 \times (6 - 2)/6 = 0.209$$
$$135 < t \le 138 \quad \Rightarrow \quad \hat{r}(t) = 0.209 \times (3 - 1)/3 = 0.139$$
$$138 < t \quad \Rightarrow \quad \hat{r}(t) = 0.139 \times (2 - 2)/2 = 0.000$$

(b)　$\mathrm{Var}(\hat{r}(100)) = 0.645^2 \times$

$$\left(\frac{1}{27(27-1)} + \frac{2}{26(26-2)} + \frac{1}{24(24-1)} + \frac{1}{22(22-1)} \right.$$
$$\left. + \frac{1}{21(21-1)} + \frac{1}{18(18-1)} + \frac{1}{17(17-1)} + \frac{1}{16(16-1)} \right)$$
$$= 0.0091931$$

The confidence interval is

$$(0.645 - 1.960 \times \sqrt{0.0091931},\, 0.645 + 1.960 \times \sqrt{0.0091931}) = (0.457, 0.833).$$